217-50
ESF
D0205708

Information in biological systems:
the role of macromolecules

Information in biological systems: the role of macromolecules

WERNER HOLZMÜLLER

Karl Marx University, Leipzig

Translated by Manfred Hecker

Université d'Ottawa
BIBLIOTHÈQUES
LIBRARIES
University of Ottawa

The right of the University of Cambridge to print and sell all manner of books was granted by Henry VIII in 1534. The University has printed and published continuously since 1584.

CAMBRIDGE UNIVERSITY PRESS

Cambridge

London New York New Rochelle

Melbourne Sydney

109355

Published by the Press Syndicate of the University of Cambridge
The Pitt Building, Trumpington Street, Cambridge CB2 1RP
32 East 57th Street, New York, NY 10022, USA
10 Stamford Road, Oakleigh, Melbourne 3166, Australia

Originally published in German as *Makromoleküle als Träger von Lebensprozessen* by Akademie-Verlag, Berlin, 1981 and © Akademie-Verlag Berlin 1981. First published in English by Cambridge University Press 1984 as *Information in biological systems: the role of macromolecules* English edition © Cambridge University Press 1984

Printed in Great Britain at the University Press, Cambridge

Library of Congress catalogue card number: 83-26331

British Library Cataloguing in Publication Data

Holzmüller, Werner
Information in biological systems: the role of macromolecules.
1. Macromolecules
I. Title II. Makromoleküle als Träger von Lebensprozessen. *English*
574.19'24 QP801.P64

ISBN 0 521 25977 0 hard covers
ISBN 0 521 27783 3 paperback

QH
507
.H6513
1984

AS

CONTENTS

PREFACE

In recent years we have experienced great advances in the chemistry, physics and technology of macromolecular materials. Their application in industry and in our households has reached an unforeseen extent. The spectrum of the technological applications of polymers ranges from plastics as constructional materials through foams, foils and elastics, i.e., rubber-like materials and synthetic fibres, to paints and adhesives. We might be tempted to consider these materials as being of exclusively technological importance and to take their further development as the subject of technological research. Adopting this view means completely ignoring the fact that macromolecules are of far greater importance for biological processes than they are with respect to technological applications. Consequently we are not at the end of a scientific development but at its very beginning and we can expect the next hundred years to be full of surprising discoveries in this field. As our fossil resources, and in particular petroleum, become exhausted, modified biomolecules, e.g., cellulose, will gain even more technological importance.

Life functions are mostly investigated by biologists using the wealth of knowledge acquired by botany and zoology as a basis. Starting from an organism as a whole they come to the cell, then the cell nucleus, then the chromosomes and finally the macromolecule. It is a fascinating idea to take the opposite direction for once, i.e., to apply the laws concerning the formation, structure and mobility of macromolecules to biomolecules, and to consider the immense variability in the composition of these molecules and the associated possibilities for storing information and fulfilling various functions. This approach may bring up some aspects which are unfamiliar to biologists but better known to polymer scientists and physicists. Above all, it is hoped that the attention of young scientists can be drawn to this promising field. The present booklet addresses chemists and physicists, with the intention of showing that with the technological application of macromolecules this research subject has by no means been exhausted.

Some facts that are well known to the biologist have been included in an easily comprehensible form so that physicists and polymer scientists will be able to follow the essential arguments. Similarly, a few fundamentals of the physics and chemistry of macromolecular substances are elucidated for biologists. Also, the areas of thermodynamics, information storage and closed-loop control, with which a physicist is quite familiar, are considered. It is not attempted to treat these topics exhaustively but to ensure that the subsequent considerations will be understood.

This booklet cannot be, and is not intended to be, a compendium of biology. Those who want to study the fundamentals of biology should refer to reputable textbooks. It is meant to contribute towards making the diverging and narrowing disciplines approach one another again. The treatment of the subject matter centres around the concept of the 'macromolecule'. An attempt is made to promote scientific research in its entirety and to check the narrow specialist approach.

Particular emphasis is placed on the peculiar characteristic of polymeric molecules of being highly variable chemically due to the combination of a few different monomeric units. The compounds thus formed are characterized by thermal and mechanical stability. The possibilities of correlating information with the structures of macromolecules are discussed extensively. Similarly structured but chemically different macromolecules reflect the multiplicity of life processes.

The growing importance of the biosciences and their penetration by the principles and methods of physics justify the publication of this monograph. Obviously, owing to the great extent of the subject it cannot be covered completely. This would require a team of authors with consequent loss of uniformity in the treatment of the material. As a nonbiologist I have ventured to write such a booklet alone and I hope it will still meet with a friendly reception.

The bias of the book is towards physics. Here the basic laws of phenomenological thermodynamics, in particular the second law, are applied to biological processes. The application of statistical thermodynamics to the storage and processing of information shows that there exists more than a formal analogy. Applying the equations of control technology to flow equilibria gives a better insight into the existence and maintenance of thermodynamic nonequilibria.

The possibilities for macromolecules constituting cybernetic systems are also discussed. Finally, comparisons are made between computers and living organisms. Life as a whole and as a hierarchic process of development is contrasted with an analytical approach. Philosophical aspects are also of importance. It will be shown that life in our world has not originated by chance. In agreement with Hoyle [49*a*] the question is discussed that life is omnipresent in the universe. Thus, no perpetual motion of the second kind is necessary to understand the origin of life. Our beautiful world must be understood as a whole, fulfilling the laws of ecology.

The content of this booklet coincides in part with some papers read by the author at meetings of the Physical Society of the GDR in the years 1970–75.

My thanks are due to the publishers for the support they have given me and to the referees for their critical remarks, which I have considered in revising the manuscript. I am also grateful to Dr. M. Hecker who has done much to make a good English translation.

SYMBOLS

A	affinity
	work
a	monomer length
	molecular chain unit
	constant in van der Waals' equation
b	molecular chain unit
	constant in van der Waals' equation
c	concentration
c_a	concentration (generalized)
c_v	specific heat at constant volume
F	free energy (work)
G	free enthalpy
H	enthalpy
	information value/content
h	Planck's constant
I	flux
	information (content)
K	constant in information theory
$K(T)$	rate constant
k	Boltzmann's constant
M_{b}	molecular weight of basic unit (monomer)
M_{p}	molecular weight of polymer
M_z	number average of molecular weight
m	mass
	number of neighbours exerting influence
N	number of macromolecules
	number of messages
	degree of polymerization

N_A	Avogadro's number
N_i	mole number
n	number of monomers in a coil
	number of microstates of a system
P	probability
P_i	number of monomers in polymer
	degree of polymerization
$\bar{P}_w$	weight average for degree of polymerization
$\bar{P}_z$	average degree of polymerization
p	pressure
	number of polyads
p_a	pressure (generalized)
Q	quantity of heat
R	gas constant
r_0	van der Waals' radius
r_0, r_1, r_{-1}	coefficient determining correcting value
S	entropy
s_0, s_1	coefficient determining control quantity
T	thermodynamic temperature
T_a	temperature (generalized)
t	time
U	potential
	nonuniformity
	internal energy
V	volume
w	desired value
X_a	generalized forces
x	control deviation/quantity
x_a	auxiliary control variable
y	correcting value
y_a	auxiliary correcting variable
Z	extensive quantity
z	disturbance
α	low-molecular weight combination of nucleotides
	information content in bits per purine or pyrimidine base
	error rate in copying
β	low-molecular weight combination of nucleotides
γ	low-molecular weight combination of nucleotides
λ	wavelength
μ_i	chemical potential of *i*th component
ν	frequency

1

Life and its molecular carriers:
a reflection

We live in a real, recognizable world. This world of ours is governed by irrefutable, inviolable laws, the laws of nature. They are an expression of the law of causality: equal preconditions for an event cause equal effects. In this world, life has developed. It has been subjected to the laws of nature without any restriction. All life processes are associated with recognizable material changes [121].

As recently as in the nineteenth century the existence of a *vis vitalis*, a force acting outside nature and controlling life, was assumed. The research work of Darwin, Wöhler, Fischer, Hill, Eigen [25] and Watson & Crick [116], to mention but a few, has shown that this concept is untenable. Moreover, ingenious statistical considerations are put forward which provide arguments for the existence of life in a random world [63]. On the other hand there are concepts which interpret life as a cosmic development that is taking place by necessity [6, 7, 8, 28]. Again, in many works much emphasis is placed on the assignment of information to material carriers and the applicability of information theory to all life processes [88, 91, 118].

Life processes are complex and manifold; the variability and adaptability of the material carriers attached to them must accordingly be great. Only complex, structured, stable macromolecules existing in innumerable variants can reflect the many-sided phenomena and forms of life. If we want to recognize these relationships we must study the life processes on the one hand and find out the corresponding changes at the molecular level on the other. The successes of the past few years speak for this approach. Even for thinking processes we find a relation between the mental content and the formation of both networks and measurable electrical potentials in the cortex [2, 4, 107, 110]. Frederick Engels pointed to this dialectic correlation in his book *The Dialectic of Nature* [29] (see also Section 10.1).

The research worker of our day resembles the discoverer of the ancient hieroglyphs and the archaeologist who collected stones covered with ancient inscriptions and deciphered them. But the biologist is faced with a more difficult job. He has to recognize the slabs with the inscriptions, i.e., the molecules carrying the information, and study their structure before he can think of deciphering them. All the time he is aware that the molecular characters, or rather, the molecular structure and composition, convey a message to him but that the content of the message, the information, has quite a different quality.

The words of Socrates 'I know that I understand nothing' are unreservedly applicable to our knowledge of the functions fulfilled by the carriers of life processes. With respect to the material structure of the correlated macromolecules we can say with some pride, but also in due modesty, that we have begun to unveil nature's secrets.

In trying to acquire knowledge the scientist works analytically. He examines all life processes; he observes simultaneously-occurring cellular changes; he employs radionuclides to explore certain molecular changes; and he gains insight into details using the electron microscope, X-ray analysis, especially small-angle scattering, spectroscopy at all wavelengths, chemical indicators and trace analyses. Some researchers, such as the Soviet scientist Engelhardt [28], consider life as a whole. They recognize the hierarchy of all living organisms and the interplay of the functions of life.

Mental functions and molecular processes are complementary to each other. Therefore, in dealing with the material carriers of spiritual relationships we shall repeatedly refer to the theory of reflection.

Living beings and life processes are analogous to computers and, in the figurative sense, they resemble the latter in their functions. The computer represents an exactly operating machine made up of electronic components. All components are interconnected and cooperate faultlessly according to the intention of the designer of the machine. The arithmetic unit with its components, such as diodes, transistors, digitizers, switches, storage and graphic displays constitutes the hardware. But programming, the language of the machine, the mental content of the addresses and the logical interplay are also constituents of the computer. It is the programmer's job to design all this; he provides the software. Without programs and algorithms a computer would be a useless device. Conversely, the most elegant and faultless programs, designed to be economical and written in the most modern programming language, are absolutely worthless without an appropriate computer. The software and the hardware belong together and make a unit. This is also true for the phenomena of life: intellectual processes, emotion and will on the one side, and the correlated material, recognizable molecular changes on the other, go together.

The term 'living being' will be used throughout this book in its widest sense. The boundary between animate and inanimate nature is very difficult to define. Usually, the transition from macromolecules to microorganisms, i.e., the formation of coacervates (precellular systems), is taken to be the lower limit, while man at his present stage of development is considered to represent the upper limit of the concept of living being. There exists a correlation between the appearance (phenotype) and the mental qualities of a living being on the one hand and a molecular arrangement (an arrangement at the molecular level) on the other. This will be discussed extensively in Chapter 10.

The observable characteristics of a living being are called its phenotype while the correlated molecular configurations realized in the cells are termed its genotype. The concept of genotype is attributed to a very definite macromolecular structure.

Besides the connection between macroscopic phenomena and their molecular reflection, which is fundamental to life processes, there is a development from simple to complex or from primitive to highly sophisticated taking place in the course of time. This evolution concerns not only the mental qualities and appearance of an organism but also their molecular reflection. We speak of a Darwinism of the molecule [47, 56].

So far, the expression 'Darwinism of the molecule' has not been used in the literature; its meaning is defined as follows. Starting from Darwinism as the theory of the phylogenetic development of living beings we associate with it a theory of a simultaneous development of the macromolecules carrying the genotypic information. We assign to the individuals characterized by a greater ability to survive, a greater complexity and a better adaptability to the environment a correspondingly greater complexity of the molecules storing their genotypic information. 'Survival of the fittest', therefore, does not only mean the survival of the individual but also the survival of the macromolecules specific to it. Just as the carriers of improved genotypic information have a better chance to survive, the same applies to the macromolecules storing the genotypic information.

Biological happenings do not resemble the mechanistic operation of clockwork but are characterized by a directionality which does not define the single events. Those who want to explore the origination and continuation of life must comprehend the interplay and complexity of life phenomena. They must not only see certain processes of motion, growth and reproduction but also understand life as a whole [28, 47]. Philosophical discussion of this point can be found in the work of Fuchs-Kittowski [33].

Scientists want to deduce comprehensive laws from the wealth of experience and, if possible, to give them a mathematical form. Thus, in physics, the laws of conservation, the symmetry principles, the laws of thermodynamics and the extremal conditions have been established as basic principles. Is there anything similar in biology? I think there is and have formulated some general biological laws. The following list does not claim to be complete.

(1) Living beings (organisms) are self-reproducing systems which change only slightly from generation to generation.

(2) In order to stabilize an individual life a multitude of controlled chemical reactions are continually taking place under isothermal and isobaric conditions in the organisms. These are closed loops (feedback reactions). I will try to show (Chapter 9) that closed-loop processes in control theory and dissipative systems in thermodynamics describe the same facts.

(3) All life phenomena, including thinking, are connected with material, recognizable processes in the cells and the nervous system.

(4) All biological processes obey the laws of nature without exception. They may be included in causal chains or start causally-related actions consciously or unconsciously.

(5) Living beings tend toward higher complexity. This is accomplished by selection and evolution. Darwin's laws apply not only to the phenotype but also to the molecular events.

(6) Living beings are in contact with their environment. The laws of thermodynamics govern these interactions.

(7) All living beings constitute a whole (system) (the concept of unity is to be expressed) and the laws of ecology control the interrelations between them.

(8) There are no immortal organisms propagating by sexual reproduction.

(9) The decisive factor for the preservation of life is energy transfer. Owing to the absorption of quantum radiation comparatively high initial energies are available at the molecular level.

(10) Mutations and coupled mutations do not occur by chance, but by necessity. The periods of time in which they take place are not determined in advance.

These postulates will be discussed in the subsequent chapters.

The principles of biological development put forward here require the existence of a number of environmental conditions which are thoroughly dealt with in my monograph *Unsere Umwelt – ihre Entwicklung und Erhaltung* (*Our Environment, its Development and Conservation*) [47]. It is there pointed out that only carbon with its great bonding versatility is a suitable basic element of organic matter. The resources of carbon, hydrogen, oxygen and nitrogen are fed into the organic cycle over and over again. Also of great importance are sulphur, phosphorus, calcium and the alkali metals.

If the elements mentioned above are combined to form low-molecular weight compounds the number of different molecules possible is of the order of only millions. This number is far too small to account for the great variety of life processes, in particular the storage of innumerable messages in the organisms. The mere requirement that all chemical processes in organisms have to take place in the temperature range 0–50 °C at atmospheric pressure in an aqueous environment implies the formation of more than 100 000 organic catalysts. Such organic catalysts are called enzymes.

The same basic processes take place in all plants and animals. Assimilation, dissimilation, reduplication (replication), protein synthesis, energy transfer, cell formation and division and the use of membranes, active and passive diffusion, and so on proceed alike in the different species. With growing differentiation, additional functions become important. These can be referred to here only in passing: metabolism, the supply of cells, an enormous extension of self-regulating processes (feedback) at all levels of living systems, regeneration, the sense organs (receptors), the locomotor system, the storage of instinctive reactions and acquired abilities, the formation of a biological equilibrium and so on. For higher organisms such as mammals, it must be assumed that over a thousand million bits of genetic information have to be stored in the germ cells. Ideally, this information should be transmitted free from errors. However, as an error is multiplied by template polymerization, error copy might lead to an unexpected change in phenotype. This can be prevented only if there is multiple storage (see Chapter 7.6).

The information must be stable. For example, it must not be destroyed by simple molecular rearrangements. This stable fixation of information can only be accomplished by the formation of definite, characteristic, primary valency bonds, bond combinations and cross-linking in macromolecules. Low-molecular weight substances are excluded because their chemical variability is insufficient. It would also be hopeless to interpret the storage of many millions of sensations in learning processes as being due to chemical bonding in small molecules. Storage by means of the formation of molecular conformations stabilized through simple dipole bonding forces and dispersion forces does not have the stability required to conserve the information.

Magnetic storage, which has a very fast response and a high capacity, is also out of the question on account of its too low stability. Life based on magnetically operating systems might be easier to realize than life forms based on silicon chemistry. This assumption is purely speculative. Living beings of such a fictitious nature are called macrobes. Organisms having magnetic storage systems, i.e., those working like technical computers, and living beings which are not based on the chemistry of carbon are quite unlikely to exist in the universe.

A certain amount of molecular mobility allowing the formation of superstructures is also of importance. The transmission of information is an essential function and is carried out most simply by means of long chainlike molecules. It is also known, from technological experience, that macromolecules make excellent constructional materials. Thus they are well suited to constituting the connective tissue in organisms. Hence, highly elastic polymeric compounds are involved in the formation of muscles and sinews. If, in the course of development, more information is added, macromolecules can be enlarged without destroying the basic structure.

In conclusion, by virtue of the necessity to satisfy principles (1) to (10) above, and from direct experimental observations, it can be inferred that macromolecules play an indispensable part in biological events. Hence, it is appropriate, after these introductory remarks, to deal with the structure, bonding variability and formation of macromolecules.

2

What are macromolecules?

Macromolecules are chemical compounds consisting of numerous atoms of the same as well as of different kinds which are linked together by strong exchange forces (covalent bonding forces). In general, they are made up of more than 100 basic molecules (monomers) corresponding to a molecular weight of over 10 000.

2.1 Bonding forces in molecules

The combination of atoms to form larger units is accomplished in some cases by ionic, i.e., Coulombic, forces but mainly by covalent interactions dependent on electron exchange. Both forces are always operative at the same time, and this is particularly obvious in, e.g., the case of the angular water molecule. In what follows the consideration will be chiefly of valence bonding forces.

Valence bonding forces have to be distinguished from molecular bonding forces, mainly van der Waals' forces, which are responsible for the formation of condensed phases, i.e., for the combination of single molecules. The forces acting between the single atoms and electrons of the same molecule are called intramolecular forces whereas those acting between neighbouring molecules or different loops of the same molecule are referred to as intermolecular forces. Repulsive forces, which increase extremely rapidly with decreasing distance between molecules and atoms, must always be taken into account. The smallest distance, r_0, that can be reached owing to the action of these repulsive forces is called the van der Waals' radius. The forces under discussion are summarized in Table 2.1.

2.1.1 *Valence bonding*

For the chemistry of macromolecules valence bonds between carbon atoms and between carbon atoms and atoms of other elements

Table 2.1. *Bonding forces in molecules*

Valence bonding		Molecular bonding	
(1)	Ionic bonding (predominantly Coulombic interaction and chemical exchange forces superimposed on it)	(1)	Dispersion forces (to be calculated quantum mechanically, taking into consideration higher orders)
(2)	Atomic bonding (predominantly exchange forces and weak Coulombic interaction superimposed on them)	(2)	Dipole bonding
		(3)	Induced dipole formation
		(4)	Hydrogen bonding
		(5)	EDA bonding (electron donor-acceptor interaction)

are of prime importance. Valence bonds can be understood only from a quantum-mechanical point of view, namely, by calculating the energy eigenvalues using the Schrödinger equation. If several possible states are characterized by the same energy they are called degenerate. We shall essentially restrict ourselves to cases where the energy levels are different and where, in equilibrium (i.e., in the bound state), configurations characterized by minimum energy are present [83, 89, 100].

The spatial distribution of the s-, p- and d-electrons is described by ψ-functions as solutions of the time-independent Schrödinger equation. Valence bonding in organic chemistry is based on σ- and/or $\sigma-\pi$-bonds. The charge distribution of the σ-bond is rotation-symmetrical with respect to the bond axis (giving free rotation about the bond) (Fig.

Fig. 2.1. Molecular orbitals; σ-bonds (*a*) between s-states and (*b*) between p_z-states (where *z* is the direction of the line joining the bonding atoms); (*c*) sp-hybridization.

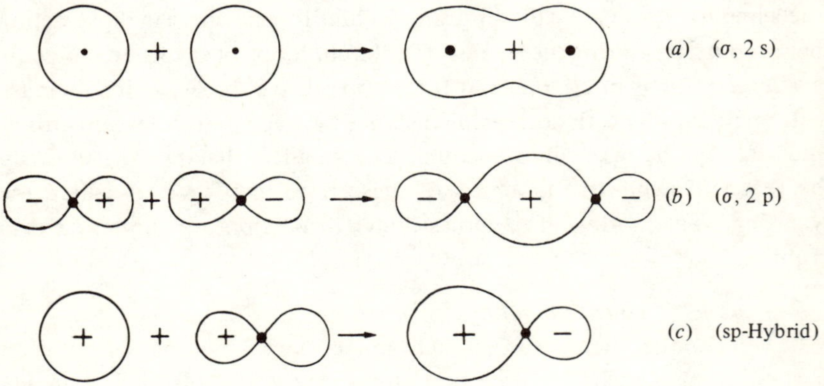

2.1 *a*, *b*, *c*). The σ-bond is caused by the overlapping of two s- or p-orbitals or of one s- and one p-orbital.

The line joining two carbon (C) atoms is defined as the z-axis. If there are p_x- or p_y-orbitals we find less overlap than is the case with the σ-orbitals in the z-direction. This is called π-bonding (Fig. 2.2). In a π-bond the charge distribution is unsymmetrical with respect to a plane through the bonding atoms (e.g., the xz-plane). As can be seen from Fig. 2.2, the rotation of orbital B about the z-axis with A fixed reduces the overlap. π-bonding leads to a stiffening of the molecule.

In many cases the electron of one bonding atom is in the s-state while that of the other atom is in the p-state (this is also true for d-states). This situation is referred to as hybridization resulting from the linear combination of the solutions of the Schrödinger equation [83, 89]. There are, for example, axially symmetrical σ-bonds described by linear combinations of s- and p-states (sp-hybrids) (Fig. 2.1 *c*).

If two p-states and one s-state participate in a single bond (proportional linear combination) the situation is referred to as sp^2 hybridization. We speak of an sp^3 hybrid if three p-states and one s-state constitute the linear combination of the Schrödinger equation to be solved. Carbon in the ground state is expected to have two 2s- and two 2p-electrons. Two s-states would saturate each other, resulting in divalent carbon. A higher bonding energy, however, is provided by tetravalent carbon, which is characterized by four unpaired electrons. We assume four equivalent sp^3-hybrid states (Fig. 2.3 *a*) and optimize the linear combinations of the four corresponding ψ-functions. The resulting eigenfunctions represent four dumbbell orbitals oriented along the sides of the tetrahedral angles as viewed from the central carbon atom. In this case, four orbitals overlap with four other orbitals of bonding s- or p-electrons. Optimum overlap is thus obtained for the resulting four σ-bonds and the observed bonding energies can be calculated e.g., for methane and its derivatives (Fig. 2.3 *a*).

Owing to their axial symmetry σ-bonds can rotate freely about the valence direction. Such a bond can also be formed by two carbon atoms,

Fig. 2.2. Overlap of p-orbitals to give π-bond (C, carbon atoms).

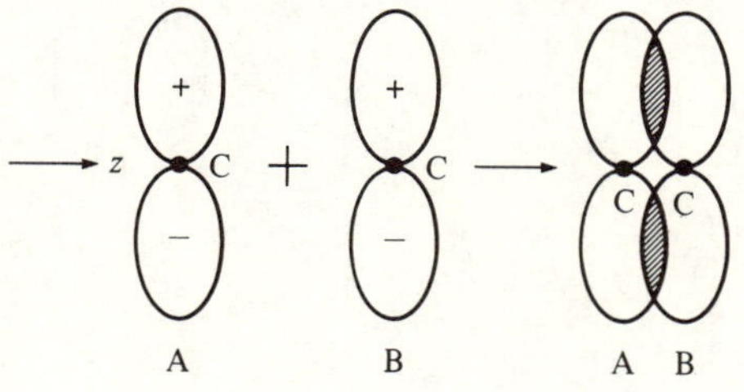

 e.g., in ethane (Fig. 2.3*b*). The carbon double bond is more difficult to describe. Here three sp^2-hybrids provide the high bonding energies that are observed. In all cases the energy eigenvalues of the Schrödinger equation are calculated using spherical functions and the observed bonding energies are obtained by linear combination. sp^2-hybridization, which is actually found in nature, forms orbitals lying in a plane and linked by one σ-bond (Fig. 2.3*c*). Two other atoms, e.g., hydrogen atoms, are connected with the carbon atoms by σ-bonds. There remains a fourth orbital for each of the two carbon atoms which is normal to the C—C direction in the plane of projection. These overlap to make a π-bond so that the C=C double bond consists of one σ- and one π-bond. The p-electrons are most likely to be located in the plane of projection. This implies hindered rotation about the C=C valence axis (Fig. 2.3*c*).

For the simple σ-bond a bonding energy of about 320 kJ/mol is observed and calculated; the σ- + π-bond has an additional amount of about 240 kJ/mol owing to the additional bond, which makes a total of approximately 560 kJ/mol. A similar consideration shows that the carbon triple bond is realized by two sp-hybrids forming a σ-bond and two p-p-linkages, i.e., π-bonds, oriented normal to the former. Roughly, this accounts for a further 240 kJ/mol, resulting in a total of about 800 kJ/mol.

The benzene ring can also be explained in terms of a linear combination of sp^2-hybrids: two of them form σ-bonds with the neighbouring carbon atoms and the third makes a σ-bond with a hydrogen atom. There are also three π-bonds whose spatial positions are not fixed but which must be taken into account. They fix the fourth electron of each carbon atom.

Fig. 2.3. Molecular orbitals and hybridizations for (*a*) methane; (*b*) ethane; (*c*) ethylene.

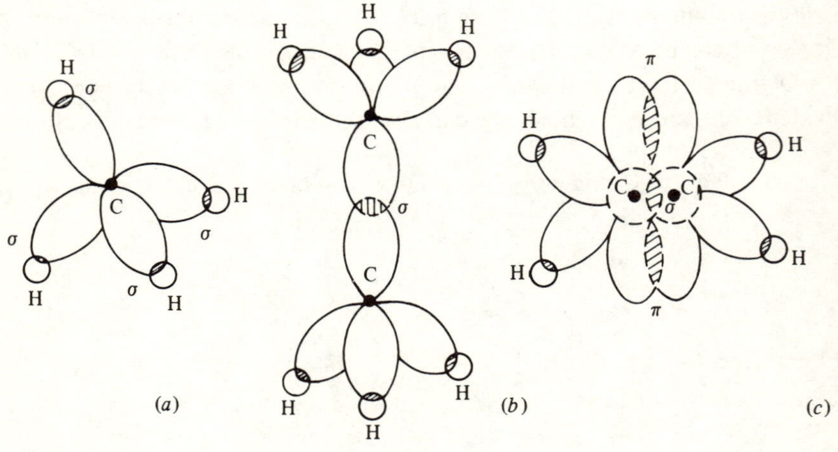

Molecular bonding forces are superimposed on the strong valence bonding forces (electron exchange forces) discussed above. The latter are strong but are effective only over a range of a few ångström units from the bonding atom. In contrast, the molecular bonding forces are much weaker (up to 40 kJ/mol) but they may be operative over longer ranges and are responsible for the formation of condensed phases [103]. There are five kinds of these weak bonding forces.

(1) *Dispersion forces.* These act between all atoms and can be calculated quantum mechanically. They are weak and have ranges of only about 10^{-1} nm. Their bonding energy relative to one mol (6.023×10^{23} bonds) is 4–12 kJ/mol. In a macromolecule all dispersion forces originating from carbon, hydrogen, oxygen, nitrogen and chlorine atoms are effective simultaneously. The bonding energies for the whole molecule in interaction with its neighbours must be added. Hence polyethylene, which has from 100 to 1000 bonds between two molecules is a solid whereas ethylene, where the same forces act only between two atoms, remains in the gaseous state even at low temperatures.

(2) *Dipole bonding forces.* Electrical dipole fields will arise if the atoms forming a molecule carry different positive and negative charges. Hydrocarbons are nonpolar while macromolecules containing oxygen generally have dipoles. The same holds true for chlorine- or nitrogen-containing polymers. Neighbouring dipoles tend to arrange themselves antiparallel and are attracted in inhomogeneous fields. Dipoles give rise to charge transfer in neighbouring neutral molecules, thus inducing dipoles (induction effect). The resulting attractive force has a range of approximately 1 nm and is frequently two to three times stronger than the dispersion forces.

(3) *Hydrogen bonding.* If in neighbouring macromolecules atoms of nitrogen and hydrogen or hydrogen and oxygen are juxtaposed there will be bonding electron pairs (*a*) as well as lone electrons (*b*) which exert a rather strong interaction on the central hydrogen atom:

$$\underset{(a)}{\mathrm{O:H}} \leftrightarrow \underset{(b)}{\mathrm{:N}}$$

Electron acceptors act as proton donors and vice versa. Together with the always existing dispersion forces the hydrogen bond (↔) provides a bonding energy of 20–40 kJ/mol. (This result was obtained taking into consideration the repulsive forces mentioned below.) These intermolecular forces play a decisive role in biomolecules and also in macromolecular materials of technological importance (such as polyurethane and polyamide (nylon)).

(4) *Electron donor-acceptor complexes.* Electron donor-acceptor (EDA) interactions occur when the parts of a molecule which tend to accept electrons are adjacent to those which donate electrons. For example, electron acceptors are halogens, nitro compounds and quinones. Electron donors are characterized by a low ionization energy. Among them are aromatic and unsaturated hydrocarbons, aniline and pyridine. The easily-induced EDA process leads to the formation of the EDA complex.

(5) *Repulsive forces.* These are active in all cases. They prevent intimate contact among atoms and determine the interatomic distances (van der Waals' radii). As molecules approach each other, e.g., as a result of the application of high pressure, the repulsive forces greatly increase.

All the forces acting at the same time constitute a potential, U, which is decisive for the strength and the structure of macromolecular bodies. Owing to thermal motion the molecules or parts of them vibrate and rotate. The observed interaction between the thermal vibrations and the potential determines strength and softening.

2.2 The structure of macromolecules [45, 58]

2.2.1 *The structure of single molecules*

Let us consider chainlike molecules with C—C or C=C bonds in solution or in the liquid phase. Besides macromolecules with molecular weights over 10 000 there are oligomers with molecular weights in the range 1000–10 000. Fig. 2.4 presents some characteristic structures of individual macromolecules. Owing to free rotation about the C—C bond, each consecutive carbon atom lies on the curved surface of a cone (Fig.

Fig. 2.4. Characteristic structures of single macromolecules in solution; (*a*) free rotation in the σ-bond; (*b*) coil formation; (*c*) cross-linking.

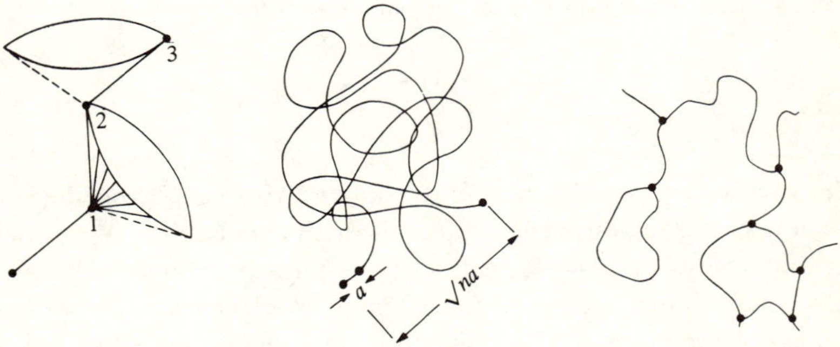

(*a*) (*b*) (*c*)

2.4*a*). Thus the third carbon atom takes a position relative to the starting-point which is difficult to determine. Then n monomeric units form a coil as shown in Fig. 2.4*b*. By monomers we understand the basic molecules, often of the same kind, that constitute the macromolecule. The average distance between the beginning and the end of the molecule (its 'diameter') is $\sqrt{na}$, where a is the length of a basic molecule. Frequently, some loops of the macromolecule are branched. This gives rise to side chains and sometimes leads to cyclization or cross-linking (Fig. 2.4*c*). In general, the length of technical macromolecules is nonuniform. If a macromolecule consists of P_i basic molecules (monomers) its degree of polymerization is P_i. If the monomer has the molecular mass M_b the molecular mass M_p of the polymer is given by

$$M_p = P_i M_b.$$

The degree of polymerization of a technical polymer is variable over a wide range. We have an average degree of polymerization

$$\bar{P}_z = \sum_i P_i n_i / \sum_i n_i$$

and a number average of the molecular weight

$$M_z = \bar{P}_z M_b.$$

n_i is the number of macromolecules of degree of polymerization P_i and $N = \sum_i n_i$ is the total number of macromolecules. In many cases it is not the number of macromolecules n_i but the mass of these molecules $M_i = n_i P_i M_b$ which determines their physical properties, e.g., in ultracentrifuge experiments. The weight average for the degree of polymerization, $\bar{P}_w$, is obtained by multiplying each degree of polymerization P_i by its mass fraction $n_i P_i M_b / \sum_i n_i P_i M_b$ and taking the sum over all P_i:

$$\bar{P}_w = \sum_i n_i P_i^2 M_b / \sum_i n_i P_i M_b.$$

Since the weight average $\bar{P}_w$ does not coincide with the number average $\bar{P}_z$ we define a nonuniformity U:

$$U = (\bar{P}_w / \bar{P}_z) - 1.$$

This is zero for a uniform molecular length because then $\bar{P}_z = \bar{P}_w$.

The variability of the different possible macromolecules arises from not only the chain length but also the various side groups that may be bound to the carbon atoms of the chain. In the simplest case all the free valencies are saturated by hydrogen atoms. This gives polyethylene:

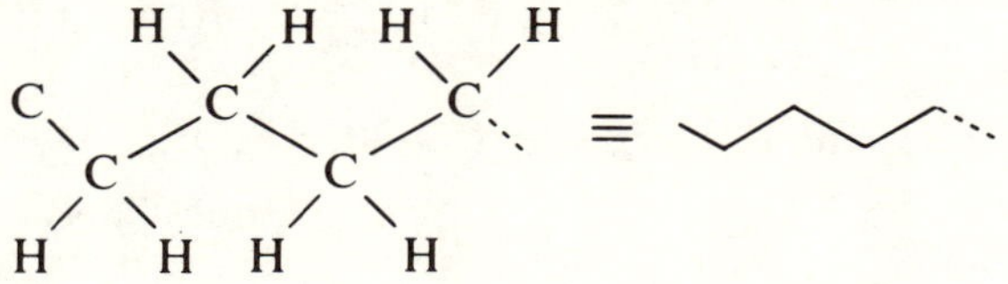

A hydrogen atom of polyethylene may be replaced by a CH_3 group in every second CH_2 group to give polypropylene:

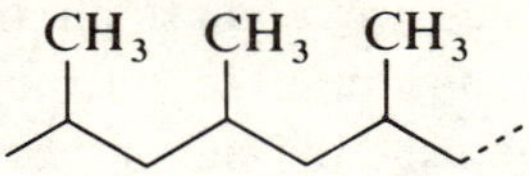

Alternatively, inserting benzene rings yields polystyrene:

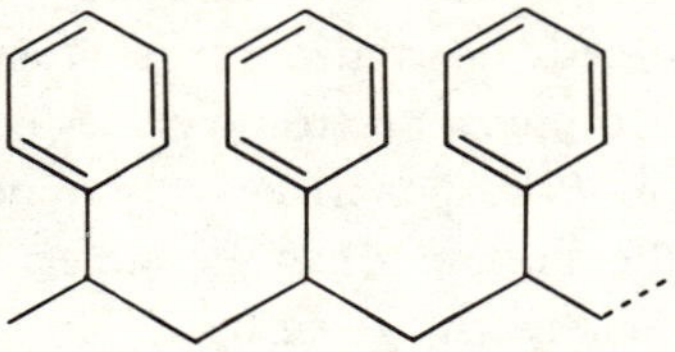

In polybutadiene every fourth chain linkage is a $\pi - \sigma$-bond, i.e., a double bond:

So far, only polymers consisting of carbon and hydrogen atoms have been mentioned. The variety becomes larger if, for example, in each of the monomer units we replace one hydrogen atom with chlorine. This gives polyvinylchloride (PVC):

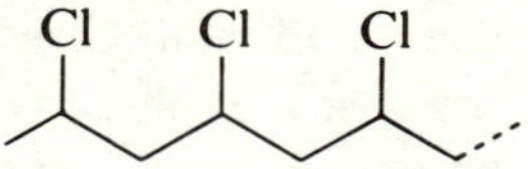

Finally, here are a few examples of oxygen- and nitrogen-containing polymers:

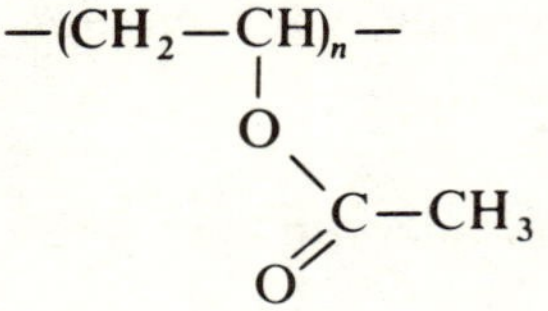

(polyvinyl acetate)

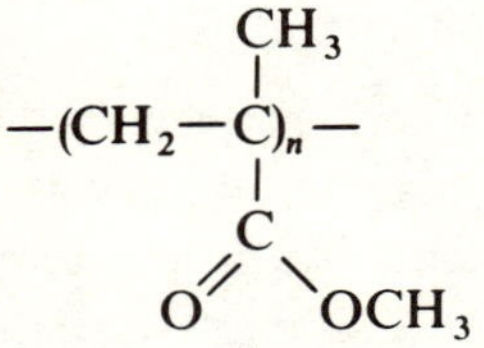

(polymethylmethacrylate)

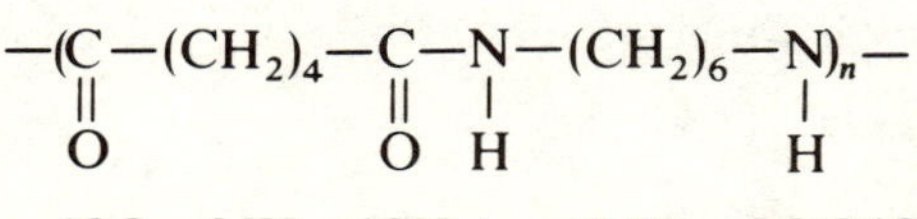

(nylon 6,6)

$$-(CO-NH-(CH_2)_m-NH-COO(CH_2)_nO)_p-$$

(linear polyurethane).

Asymmetric basic molecules can be linked in different ways to form macromolecules. Let us take as an example the vinyl chloride molecule. It may be formed by an alternating arrangement of one CH_2 and one CHCl group at a time (head-to-tail product) or of two CH_2 and two CHCl groups at a time (head-to-head product (not shown)).

Even the head-to-tail product has several stereochemical forms. If, as one proceeds along the molecule from carbon atom to carbon atom one finds the next chlorine atom always to the right of the axis of the molecule (which is a consequence of the tetrahedral angle of 109°), the configuration is isotactic:

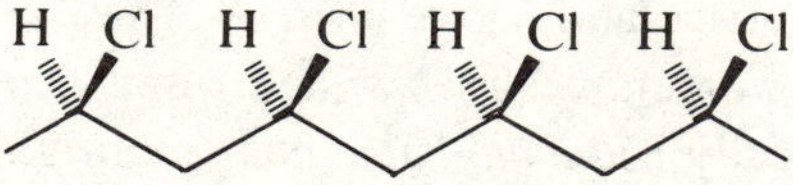

If, however, the chlorine atoms appear alternately to the left and to the right as one proceeds along the molecular axis the configuration is called syndiotactic:

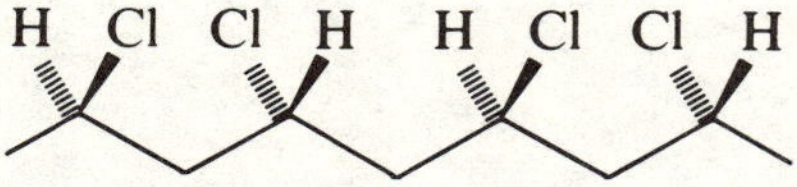

A macromolecule containing carbon $\pi-\sigma$-bonds has two isomeric forms, the *cis*- and the *trans*- forms, owing to the lack of free rotation about the C=C linkage:

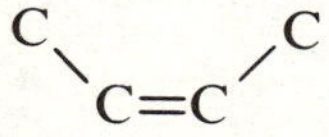

cis-form

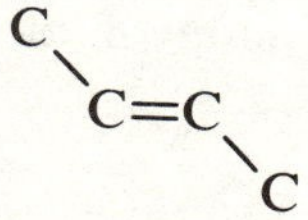

trans-form

We shall see later that the different configurations of molecules may serve to store information.

In general, a molecule can rotate about the C—C bond without changing its stereospecific structure (the isotactic and syndiotactic configurations remain unchanged). Only the conformation may be changed:

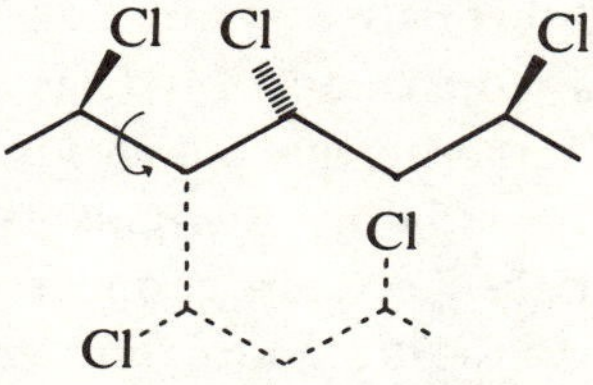

(conformation i)

(conformation ii)

Rotation in the direction of the arrow changes conformation i into conformation ii.

There is an essential difference between configurations and conformations. Configurations cannot be changed into each other without rupture of a bond. Therefore, configurations are fundamentally different from one another chemically. They are suited for the storage of indestructible information. Since the arrangement of the hydrogen and chlorine atoms in the above-mentioned molecules can be varied almost without any restriction to create a great variety of new configurations it is obvious that an immense number of chemically different molecules can be formed using the three elements carbon, chlorine and hydrogen.

In addition, there are macromolecules which are made up of different basic molecules such as ethylene and vinyl acetate. They are called copolymers. The different monomers *a* and *b* which form a copolymer are arranged either stochastically (*aaabbabbabaabbb*) or in alternating sequences (*ababab*) or else they display a block structure (*aaaaaabbbbbbb*). Also, monomers can be grafted onto macromolecules, giving graft copolymers:

```
aaaabb bbbabaab
      c
      c
      c
      c
```

As there are so many possibilities for creating chemically different molecules having one and the same empirical formula the latter may involve a large number of isomeric molecules. Even simple $\pi - \sigma \leftrightarrow \sigma$ exchanges, branching and permutations of different monomers give rise to new isomers.

2.2.2 *The structure of macromolecular solids*

By the structure of macromolecular systems I mean the spatial arrangement of the monomeric units in the molecule or of the macromolecules in the whole body. Structure is caused by the action of atomic, intramolecular and intermolecular forces. In low-molecular weight substances spherically symmetric forces, e.g., Coulombic attraction, give rise to the closest spherical packing, i.e., to cubic or hexagonal crystals. If valence bonds are involved there arise, in addition, directed forces which result in crystallization in more complex systems. Thus, in the simplest case, polymer crystals are orthorhombic but mostly they are monoclinic. The structure is preformed during the polymerization. Because of the exothermic nature of the polymerization process the macromolecule originates in the liquid phase, in a state of greater thermal energy. Therefore, at the very beginning the molecule grows as a random coil. The single loops arrange themselves parallel to one another on giving off the heat of polymerization to the surroundings of the growing molecule. Since the original loops formed during polymerization prevent a completely parallel arrangement of the individual strands without distortion of the bond angles a polymer is usually a partially crystalline body. In many cases the regions of parallel arrangement of segments are so small that they cannot be proved to be crystalline even by means of X-ray analysis.

In inorganic matter the interaction between internal energy and entropy alone determines the structure. Thus, with spherically symmetric forces a particle will occupy the point of intersection of plane or space diagonals. Molecules of regular and symmetric structure tend to crystallize even if there is only the weak dispersion bond present. This is true, e.g., for polyethylene. Irregularly-structured molecules will form coils and unordered clusters even if strong dipole forces are present. They are almost amorphous (e.g., polyvinyl chloride). If the intermolecular interactions are still stronger then even unsymmetric molecules will partly assume a parallel arrangement, which gives rise to crystallization and results in partially crystalline structures (Fig. 2.5*a*). In some cases this structure is replaced by a lamellar arrangement of crystallites, especially in polyethylene (Fig. 2.5*b*). Here the meandering macromolecules form lamellae of about 150 Å thickness [58, 104]. These lamellae are superimposed on one another like plates (Fig. 2.6*a*).

Fig. 2.5. The packing of single molecules in the formation of amorphous solids: (*a*) partially crystalline fringe structures; (*b*) lamellar structures.

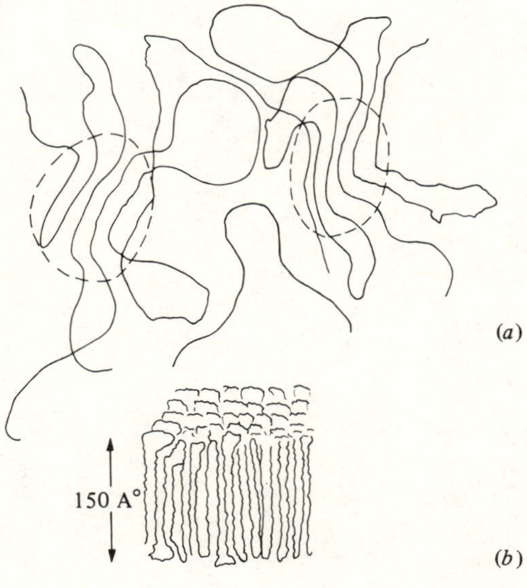

Primary crystallites frequently constitute secondary structures, such as spherulites, which are spherically symmetric (Fig. 2.6*b*). These spherulites can be made clearly visible using polarized light and can be observed to melt on heating. The formation of secondary structures may be enhanced by hydrogen bonds if there are any present. (This does not, however, apply to polyethylene.)

The phase transition and the formation of structure are determined by the thermodynamic potentials

$$F = U - TS \quad \text{or} \quad G = H - TS,$$

where F is the free energy, U is the internal energy, T is the thermodynamic temperature, S is the entropy, G is the free enthalpy and H is the enthalpy. The quantities F and G tend toward minimum values in equilibrium. In thermodynamic equilibrium, with $\mathrm{d}F = 0$, we have

$$T = \mathrm{d}U/\mathrm{d}S.$$

At low temperatures U determines the structure, a parallel arrangement of the molecules being preferred. With increasing temperature the influence of the entropy term rises and this favours the formation of amorphous regions. Of course, the above formulae are valid only for

Fig. 2.6. (*a*) Helical arrangement of lamellae in crystallized polyethylene; (*b*) arrangement of molecules in a spherulite.

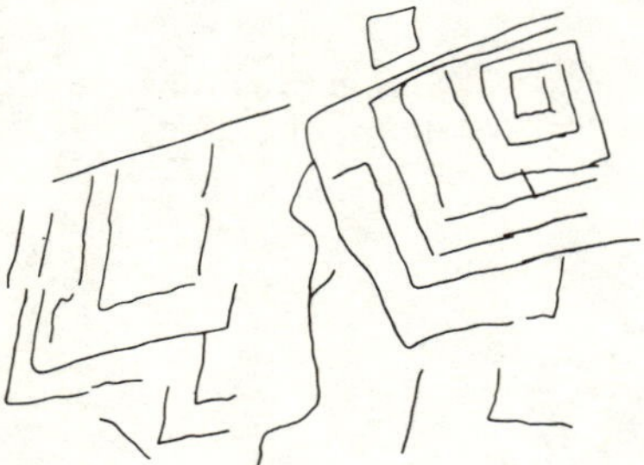

(*a*)

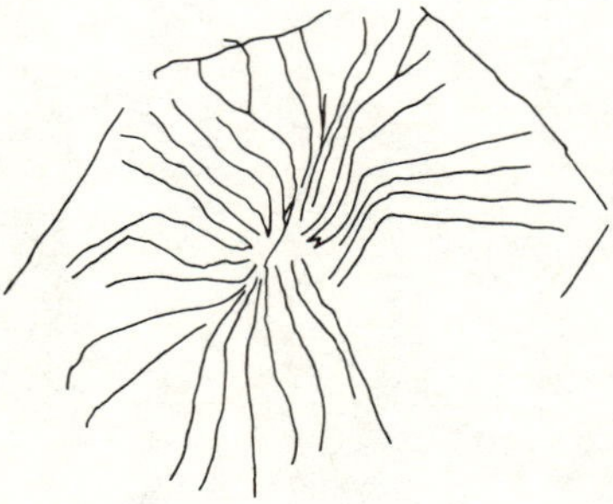

(*b*)

thermodynamic equilibrium. Solid polymers often require long annealing before this equilibrium is reached. In the softening region the crystallites melt. As the material is heated molecular cavities arise when a certain temperature, T_g, is reached. On reaching this glass transition temperature molecular mobility is stimulated such that different conformations (i↔ii) may be assumed depending on the originating holes (Fig. 2.7). By the conformation we understand a definite structure of the molecule characterized by a minimum of potential energy. At high temperatures there are fluctuations between different molecular arrangements (conformations).

If an external tensile stress acts upon a molecule the possible conformation that lies in the direction of this force will be preferred. Molecular orientation effects leading to a stretching of the material or to a rubberlike deformation are observed (Fig. 2.8). These orientations result in an anisotropy of the physical properties. The orientation is characterized by a considerable increase in molecular strength in the direction of the axis of the molecule. Crystallization also improves the mechanical properties. The same effect is achieved by cross-linking, i.e., the formation of

Fig. 2.7. (*a*) Possible molecular conformations, i and ii, when a macromolecular solid is melting. (*b*) Section of an energy hyperplane for a i–ii transition in the direction AB (ΔU_i, ΔU_k, activation energies).

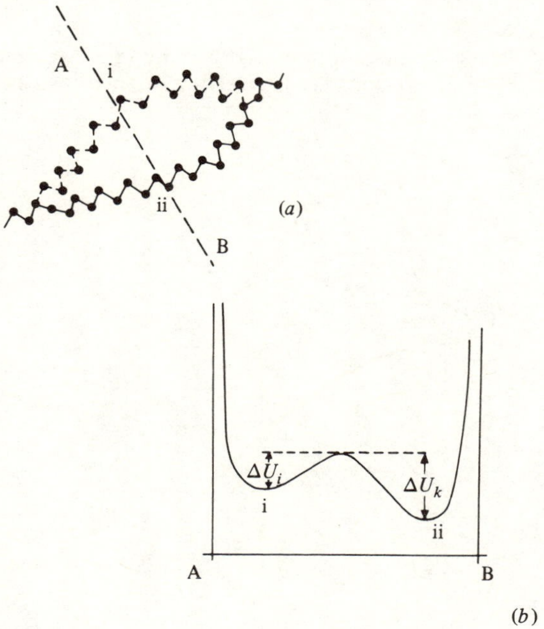

 chemical bridges between macromolecules. Such cross-linking may arise at trifunctional carbon atoms or be due to sulphur bridges (e.g., in vulcanized rubber) or hydrogen atoms between oxygen and nitrogen atoms. Cross-linking can also be accounted for by an accumulation of intermolecular bonds in crystalline regions.

On account of the simultaneous action of intermolecular bonding forces of different ranges, macromolecular solids have no sharp melting point but a broad softening region. In the temperature range between 100 and 200 K only the dispersion forces are equivalent to their thermal energies of vibration.

Solid polymers which are formed only by dispersion forces already have partial mobility at −70 °C (e.g., rubber). If dipole forces are involved the softening region (glass transition temperature) is shifted to 100 °C (e.g., PVC, polymethacrylate). Finally, substances with hydrogen bonds retain their thermal stability up to temperatures above 100 °C (e.g., polyamides, cellulose). In all cases, however, a molecular displacement or dislocation requires that some inter- or intramolecular forces have to be overcome. Since macromolecular bodies have no free electrons their electric conductivity is weak: they are good insulators. In metals free electrons are responsible for thermal conductivity (Wiedemann-Franz law). The poor thermal conductivity or the good thermal insulation of polymers is due to the absence of an electron gas.

Fig. 2.8. Molecule orientation in partially crystalline macromolecular solids (the crystalline regions are enclosed by dashed lines).

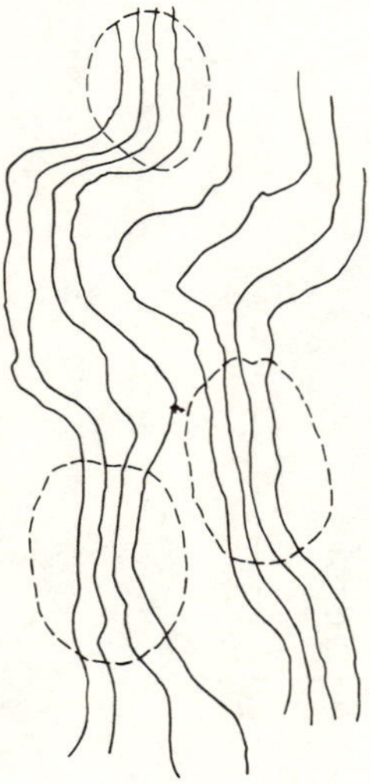

In general, macromolecules arise from basic molecules (monomers) by polymerization. What is important in this process is that C=C double bonds ($\pi-\sigma$-bonds) can be changed via radical or ionic states into single bonds (σ-bonds) and that more monomers can be attached to the growing molecule (chain reaction).

The radicals necessary for polymerization are formed by the thermal decomposition of peroxides, by irradiation or by chemically-initiated radical formation. Radicals are neutral molecules with uncompensated spin states (unsaturated valence bonds). Instead of electron pairs they contain 'unpaired' ('lone') electrons. Radical states can be illustrated by removing an atom, e.g., a hydrogen atom including its electron, from an ordinary organic molecule.

During polymerization the change from carbon π-σ-bonds to σ-bonds is decisive. One $\pi-\sigma$-bond gives rise to two σ-bonds when the monomer links on to the growing molecule. At first an activation energy of about 560 kJ/mol is required to break up a $\pi-\sigma$-bond. The two resulting σ-bonds, however, release 2×320 kJ/mol so that altogether 80 kJ/mol are given off to the environment (1 kcal = 4.186 kJ). Thus, polymerization is an exothermic process. The conversion per unit time can be considerably increased by the use of catalysts.

Let us again take polyvinyl chloride as an example. I shall use dots to represent bonding electrons, following the antiquated octet theory of Lewis. There are three steps in the process of polymerization. The first step is chain initiation. This takes place via an initiator ($R_3C\cdot$):

$$\begin{array}{ccccccccccc}
 & \mathrm{R} & & \mathrm{H} & & \mathrm{Cl} & & & \mathrm{R} & \mathrm{H} & \mathrm{Cl}\\
\mathrm{R{:}} & \ddot{\mathrm{C}}\cdot & + & \ddot{\mathrm{C}} & {::} & \ddot{\mathrm{C}} & \rightarrow & \mathrm{R{:}} & \ddot{\mathrm{C}}{:} & \ddot{\mathrm{C}}{:} & \ddot{\mathrm{C}}\cdot\\
 & \ddot{\mathrm{R}} & & \ddot{\mathrm{H}} & & \ddot{\mathrm{H}} & & & \ddot{\mathrm{R}} & \ddot{\mathrm{H}} & \ddot{\mathrm{H}}
\end{array}$$

The second step is chain propagation:

$$\begin{array}{ccccccccccccccc}
 & \mathrm{R} & \mathrm{H} & \mathrm{Cl} & & \mathrm{H} & & \mathrm{H} & & & \mathrm{R} & \mathrm{H} & \mathrm{Cl} & \mathrm{H} & \mathrm{H}\\
\mathrm{R{:}} & \ddot{\mathrm{C}}{:} & \ddot{\mathrm{C}}{:} & \ddot{\mathrm{C}}\cdot & + & \ddot{\mathrm{C}} & {::} & \ddot{\mathrm{C}} & \rightarrow & \mathrm{R{:}} & \ddot{\mathrm{C}}{:} & \ddot{\mathrm{C}}{:} & \ddot{\mathrm{C}}{:} & \ddot{\mathrm{C}}{:} & \ddot{\mathrm{C}}\cdot \quad \text{etc.}\\
 & \ddot{\mathrm{R}} & \ddot{\mathrm{H}} & \ddot{\mathrm{H}} & & \ddot{\mathrm{H}} & & \ddot{\mathrm{Cl}} & & & \ddot{\mathrm{R}} & \ddot{\mathrm{H}} & \ddot{\mathrm{H}} & \ddot{\mathrm{H}} & \ddot{\mathrm{Cl}}
\end{array}$$

The final step is chain termination:

$$\begin{array}{ccccccccccccccccccc}
 & \mathrm{H} & \mathrm{H} & \mathrm{Cl} & \mathrm{H} & & & \mathrm{H} & \mathrm{H} & \mathrm{H} & \mathrm{Cl} & & & \mathrm{H} & \mathrm{H} & \mathrm{Cl} & \mathrm{H} & \mathrm{Cl} & \mathrm{H}\;\mathrm{H}\;\mathrm{H}\\
\mathrm{H{:}} & \ddot{\mathrm{C}}{:} & \ddot{\mathrm{C}}{:} & \ddot{\mathrm{C}}{:} & \mathrm{C}\cdot & + & \mathrm{H{:}} & \ddot{\mathrm{C}}{:} & \ddot{\mathrm{C}}{:} & \ddot{\mathrm{C}}{:} & \ddot{\mathrm{C}}\cdot & \rightarrow & \mathrm{H{:}} & \ddot{\mathrm{C}}{:} & \ddot{\mathrm{C}}{:} & \ddot{\mathrm{C}}{:} & \ddot{\mathrm{C}}{:} & \ddot{\mathrm{C}}{:}\ddot{\mathrm{C}}{:}\ddot{\mathrm{C}}{:}\ddot{\mathrm{C}}{:}\mathrm{H}\\
 & \ddot{\mathrm{Cl}} & \ddot{\mathrm{H}} & \ddot{\mathrm{H}} & \ddot{\mathrm{H}} & & & \ddot{\mathrm{H}} & \ddot{\mathrm{Cl}} & \ddot{\mathrm{H}} & \ddot{\mathrm{H}} & & & \ddot{\mathrm{Cl}} & \ddot{\mathrm{H}} & \ddot{\mathrm{H}} & \ddot{\mathrm{H}} & \ddot{\mathrm{H}} & \ddot{\mathrm{H}}\;\ddot{\mathrm{Cl}}\;\ddot{\mathrm{H}}
\end{array}$$

The termination of molecular growth may occur through the addition of a hydrogen atom, through the random contact of two growing chains, through chain transfer, or through the simultaneous formation of $\pi-\sigma$-bonds with the expulsion of hydrogen atoms.

As mentioned above, ionic reactions can also produce macromolecules. They involve the displacement of electrons in connection with the formation of charged particles, e.g.:

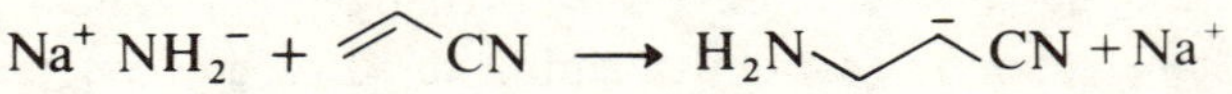

More monomers link up with the growing molecule and the negative charge may be transferred to the following group (anionic polymerization).

In other cases the formation of macromolecules is connected with the release of low-molecular weight components such as water. Such processes are called polycondensation processes, e.g.:

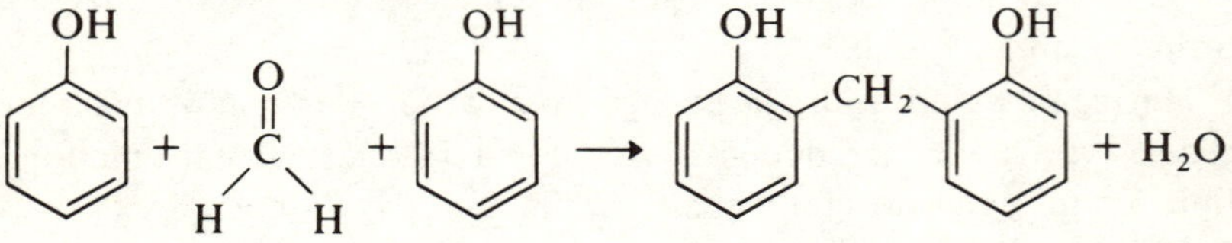

Polymers can also be formed by addition reactions, e.g., of the isocyanate group with polyhydric alcohols to form polyurethanes. The chainlike polymers resulting from such bifunctional monomers in the usual modes of preparation may be branched or cross-linked; the addition of groups such as divinyl benzene leads to spatial cross-linking.

In living systems (organisms) the macromolecules are generated by template polymerization. This gives rise to identical molecules, and will be dealt with in detail later. Recently, in the nonbiological sphere, attempts to link vinyl alcohol chains to polypyrrolidine chains through hydrogen bridges by the process of template polymerization have been successful.

3

Biopolymers

(*General references for this chapter are* [1, 34, 40, 44, 56, 87].)

3.1 Macromolecules in organisms

Living beings are made up of solid, liquid and gaseous parts. Chemical processes in organisms take place in the aqueous phase. The proportion of water in living organisms may be more than 80%.

Low-molecular weight substances are the solvents of living organisms, the ionic constituents (e.g., K^+, Na^+ and Cl^-), the energy carriers (e.g., glucose, fats, adenosine triphosphate (ATP) and adenosine diphosphate (ADP)) and the basic monomeric components of the macromolecules. There are also low-molecular weight organic acids, bases, esters, etc., substances for the transfer of 'chemical information' (hormones) and vitamins.

Polymeric substances are the basic units of which the skin, the hair, the supporting apparatus, fibres, sinews and so on are made up (frequently with low-molecular weight substances inserted). The most important polymeric components are nucleic acids, which carry the information necessary for life processes and give instructions for the biological processes, and proteins, especially in the form of enzymes, i.e., organic catalysts. Polymeric nucleic acids and proteins also provide the material basis for receptors of the transmitters of information and carriers of the mental processes, i.e., the apparatus of the nervous system.

Organisms never absorb macromolecules directly from the environment. First, macromolecules that are constituents of an organism's food must be depolymerized, this process being the most important task of digestion. Then there must be a synthesis of body-specific substances.

Some constituent polymers of cells are crystalline (e.g., cellulose); others are amorphous (e.g., starch) or have individual-specific structures (e.g., nucleic acids and proteins). In the biomass (the mass of all living matter taken together) cellulose plays the greatest role.

Owing to their interactions with water and to the chemical processes taking place in an aqueous environment most biopolymers are hydrophilic and partly soluble in water. In an aqueous environment the hydrogen bridges between polymers are not very effective. Hydrogen bonds are stronger in the absence of water.

Most biopolymers arise from the polymerization of certain monomers. Consider these examples:

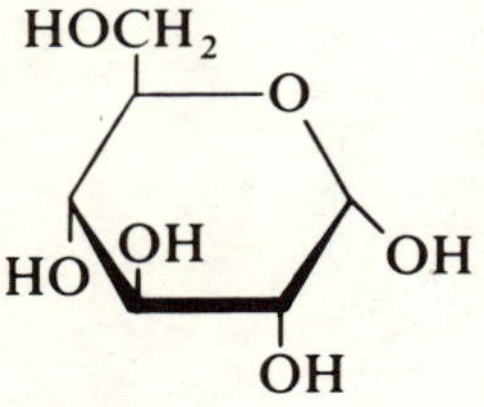

α-glucose gives starch (isotactic linkage) (3.1)

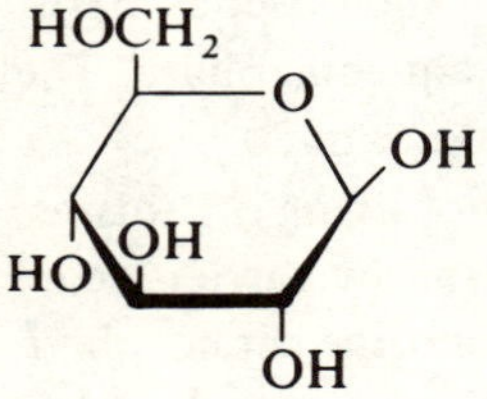

β-glucose gives cellulose (syndiotactic linkage) (cf. Fig. 4.1) (3.2)

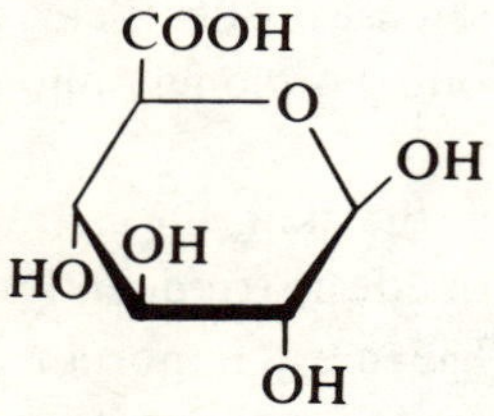

Galacturonic acid gives protopectin (constituent of cell walls) (3.3)

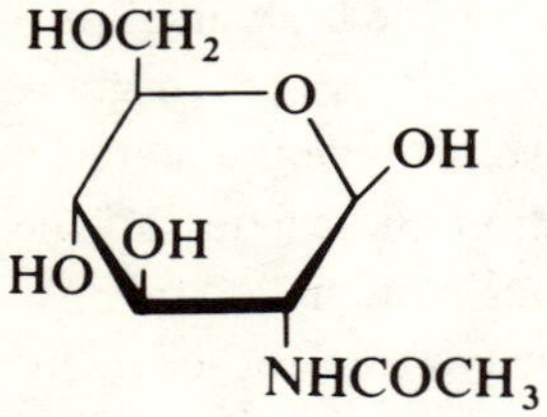

β-*N*-acetyl glucosamine gives chitin (3.4)

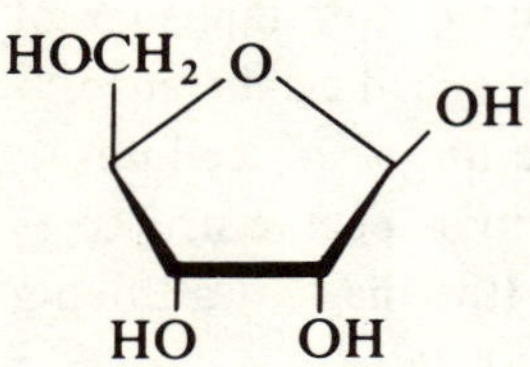

β-D-ribose, when linked with a purine or pyrimidine base gives a nucleoside; the addition of phosphate to the nucleoside yields a nucleotide (DNA component) (3.5)

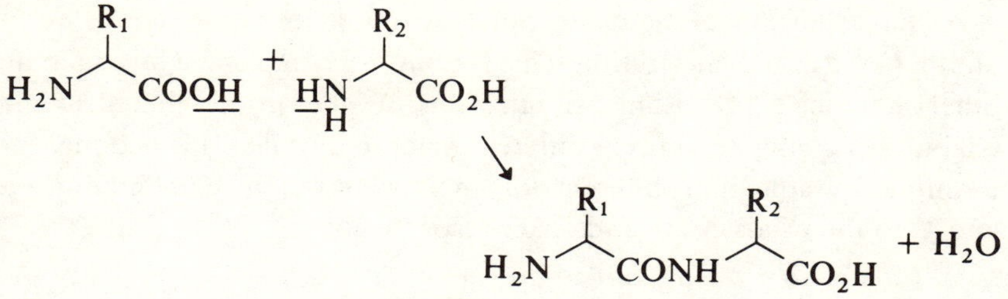

amino acids give peptides and water (3.6)

(In (3.6) water formation takes place at the underlined groups.)

3.2 The formation of structure in polymeric molecules by template polymerization

While the technical macromolecules considered in Chapter 2 are of different sizes and vary with respect to isotactic and syndiotactic structures or *trans*- and *cis*- forms owing to the random character of initiation, propagation and termination, the template polymerization of biopolymers yields an exact reproduction of existing macromolecules. This is accomplished by an almost error-free addition of low-molecular weight units to one macromolecule that is complementary to another acting as template. The average number of errors is estimated on a statistical basis in [25] and [121].

Template polymerization provides identically equal molecules in a manner which is comparable to printing the characters of a text by a printing machine. Let a and b be the units of the chain and A, C, G and T complementary side groups such that A → T, C → G, G → C and T → A (see Section 3.3). From the template

$$\begin{array}{cccccccccccccccc} a & - & b & - & a & - & b & - & a & - & b & - & a & - & b & \text{(i)} \\ & & | & & & & | & & & & | & & & & | & \\ & & \mathrm{A} & & & & \mathrm{C} & & & & \mathrm{G} & & & & \mathrm{T} & \end{array} \qquad (3.7)$$

there arises at first the template

$$\begin{array}{cccccccccccccccc} & & \mathrm{T} & & & & \mathrm{G} & & & & \mathrm{C} & & & & \mathrm{A} & \\ & & | & & & & | & & & & | & & & & | & \\ a & - & b & - & a & - & b & - & a & - & b & - & a & - & b & - \ \text{(ii)} \end{array}$$

and finally, by repeating the process, we get a molecule which is identically equal to the original one:

$$\begin{array}{cccccccccccccccc} a & - & b & - & a & - & b & - & a & - & b & - & a & - & b & - \ \text{(ii)} \\ & & | & & & & | & & & & | & & & & | & \\ & & \mathrm{T} & & & & \mathrm{G} & & & & \mathrm{C} & & & & \mathrm{A} & \\ \\ & & \mathrm{A} & & & & \mathrm{C} & & & & \mathrm{G} & & & & \mathrm{T} & \\ & & | & & & & | & & & & | & & & & | & \\ a & - & b & - & a & - & b & - & a & - & b & - & a & - & b & - \ \text{(i)} \end{array} \qquad (3.8)$$

As an extension of template polymerization it is possible to form another kind of polymerization. This variant is accomplished by attaching other monomers to certain combinations of side groups linked to the chain, thus giving an entirely different macromolecule. This is done, for example, by attaching different low-molecular weight combinations α and β to the triads ACG and AAG respectively:

$$\begin{array}{cccccccccccc} a- & b- & a- & b- & a- & b- & a- & b- & a- & b- & a- & b- \\ & | & & | & & | & & | & & | & & | \\ & \mathrm{A} & & \mathrm{C} & & \mathrm{G} & & \mathrm{A} & & \mathrm{A} & & \mathrm{G} \end{array} \quad \underbrace{\mathrm{A\ C\ G}}_{\alpha}\ \underbrace{\mathrm{A\ A\ G}}_{\beta} \qquad (3.9)$$

This formation of entirely different polymers, which is stimulated by several side groups (α, β, γ...) of a molecule is called translation or, more correctly, translation polymerization.

3.3 Nucleic acids and their formation by template polymerization [18, 61]

Template polymerization as described above is characteristic of biological processes. Let us consider it by taking the ribonucleic acids as an example. Nucleic acids are high-molecular weight compounds consisting of pentoses (5-C sugars), organic bases and phosphoric acid. Depending on the chemical constitution of the pentose present we distinguish between ribonucleic acid (RNA) and deoxyribonucleic acid (DNA). RNA and DNA contain ribose and deoxyribose, respectively. The formulae are as follows:

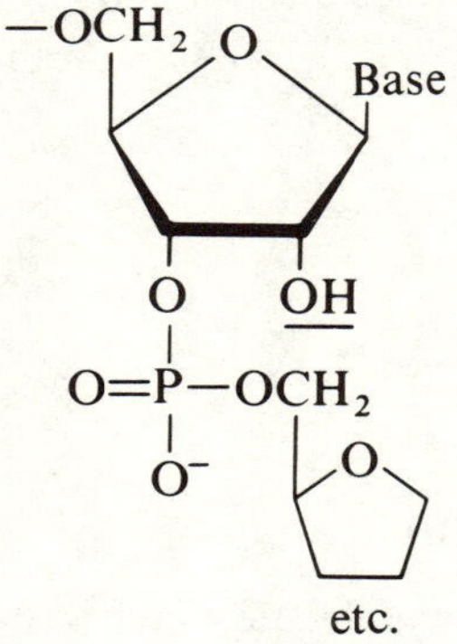

(3.10)

RNA (OH group of ribose underlined)

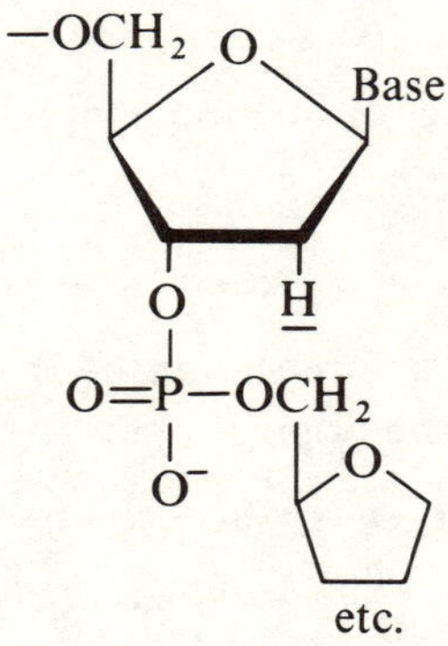

(3.11)

DNA (H atom of deoxyribose underlined)

The most important bases are:

(1) The pyrimidine bases: cytosine (C), thymine (T) and uracil (U).

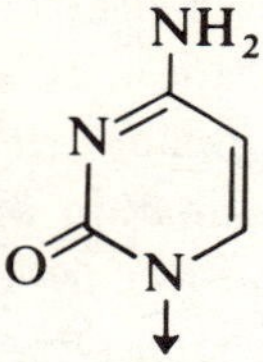
(3.12)

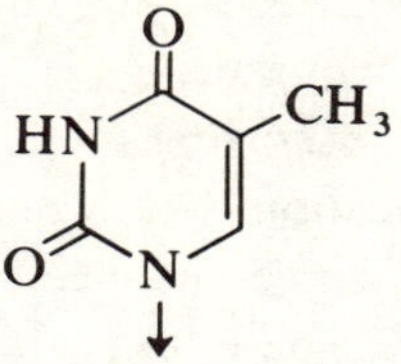
(3.13)

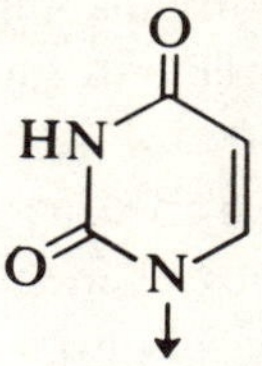
(3.14)

Cytosine (C) Thymine (T) Uracil (U)

(2) The purine bases: adenine (A) and guanine (G).

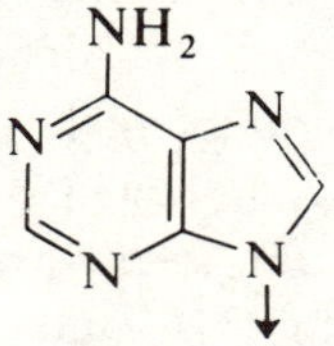
(3.15)

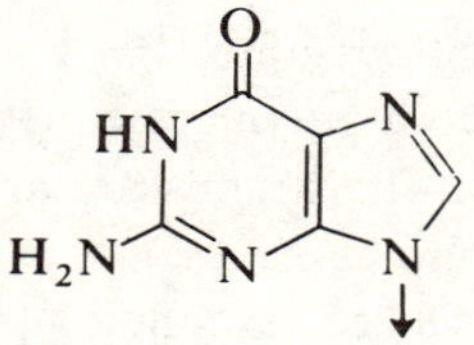
(3.16)

Adenine (A) Guanine (G)

The arrows indicate the places where the bases are connected with the ribose (or deoxyribose). The chemical combination of a base with a pentose gives a nucleoside with the elimination of water, e.g., ribose + adenine → adenosine. In almost all cases uracil is observed in RNA. In DNA, thymine is found in its place. The chemical combination of phosphate with a nucleoside gives a nucleotide.

The template polymerization of ribonucleic acids gives rise to long-chain molecules of the form

```
phosphate—ribose—phosphate—ribose—phosphate—
            |                 |
          purine            purine
            or                or
        pyrimidine        pyrimidine
           base              base
```

Which base is bound in each case is determined by the principle of template polymerization. The process takes place according to formulae (3.7) and (3.8) where a is the phosphate and b the ribose (or deoxyribose). The bases A, C, G, T and U have steric structures such that they fit in with one another when the chain molecules lie parallel (lock-and-key relation of the complementary base pairs C and G or A and T or A and U). These linkages are formed by hydrogen bonds. The nucleotides carrying different purine or pyrimidine bases are linked to the macromolecular chain by primary valence bonds to form different sequences.

 Thus, chemically different, stable molecules originate. We shall come back to the great variety of configurations thus made possible.

The continual reproduction of identically equal polymeric nucleic acids taking place in the cells of all living beings is termed replication. It is one of the most vital processes occurring in all plant and animal cells and even in coacervates. Replication makes possible the formation of molecules of identical size, the same chemical structure and the same sequence of purine and pyrimidine bases. As will be discussed later DNA is present in cells in a double-stranded form, the two strands forming characteristic helices. Hence, replication requires untwining of the strands, which results in a Y-shaped structure (Fig. 3.1). In the cytoplasm the nucleotides approach the opened double strand S_1. Aided by the enzyme DNA polymerase they form two identical double-helical strands S_2 and S_3. The strands are arranged so that the complementary purine and pyrimidine bases are exactly opposed and the steric structure is restored. The corresponding pairs are A–T (A–U in RNA) and C–G. The base sequence is free from overlap and unambiguous. It was a great scientific success when replication was achieved outside living cells. In 1958, scientists succeeded in polymerizing the nucleotides to make DNA using the enzyme discovered by Kornberg [61]. They used DNA molecules as the templates and the triphosphates of all four nucleotides as their source of monomers. In such a system the chain grows link by link with the elimination of pyrophosphate.

Fig. 3.1. Y-shaped structure of DNA strands in replication. S_1, opened double strand to be replicated; S_2, S_3, identical growing double strands.

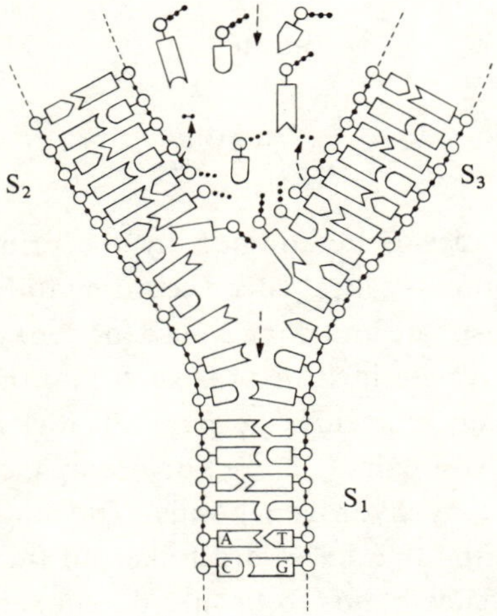

Nucleic acids (including the purine bases A and G and the pyrimidine bases C, T and U) occur in organisms as oligomers and polymers with a degree of polymerization of up to 10 000 and more. The length of single molecules can exceed 1 mm. Some nucleic acids are soluble in water. DNA appears chiefly in the cell nucleus while RNA is found in the cytoplasm. The significance and structural patterns of DNA are so well known that we need not go into more detail here [see 1, 2, 3, 35, 56, 64, 112*b*, 121].

In later chapters on molecules as carriers of life, however, we shall need some of the basic facts to understand the correlations between chemical composition or sequential structure on the one hand and information content on the other.

3.4 Proteins and their formation from amino acids [1, 40, 44, 56, 102, 112*b*, 121]

Proteins are polycondensates of amino acids, which have the formula

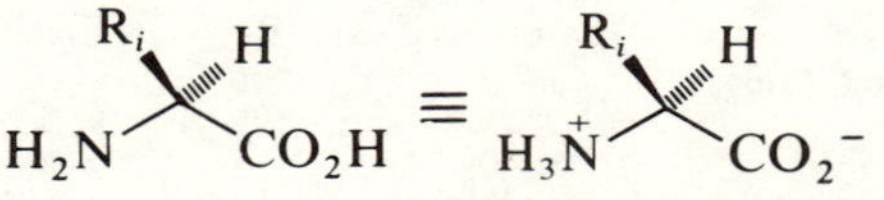

(3.17)

Zwitterion

Amino acids contain one carboxyl and one amino group as well as one side-chain residue R_i. The carbon atom linked with the residue R_i (α-carbon) has an asymmetric character except when R_i is hydrogen. Hence, amino acids may be dextrorotatory (D) or laevorotatory (L) with respect to linearly polarized light. We observe laevorotatory forms among the natural proteins.

During condensation a carboxyl group forms a peptide linkage with an amino group and water is set free:

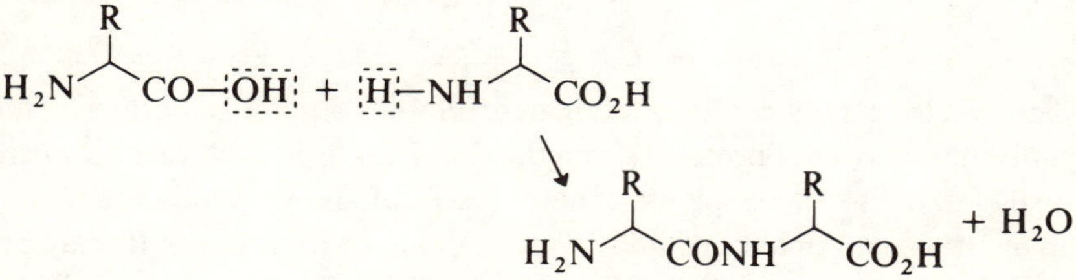

(3.18)

As the condensation process continues the polymeric peptide chain is formed:

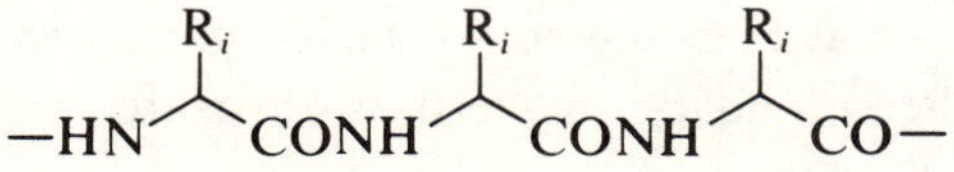

(3.19)

 Table 3.1. *Structures of the side groups of the amino acids*

Neutral

Glycine (Gly) $R_i = H$

Alanine (Ala) $R_i = CH_3$

Valine (Val) $R_i = CH{<}^{CH_3}_{CH_3}$

Leucine (Leu) $R_i = CH_2CH{<}^{CH_3}_{CH_3}$

Isoleucine (Ile) $R_i = \underset{|\;CH_3}{C}HCH_2CH_3$

Phenylalanine (Phe)

$R_i = CH_2$– 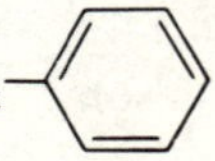

Proline (Pro) 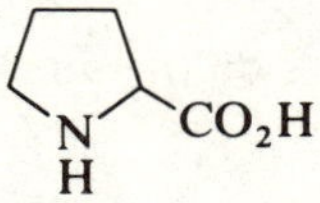

Tryptophan (Trp)

$R_i =$ 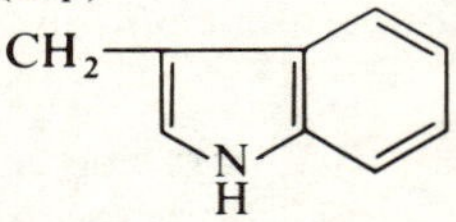

Serine (Ser) $R_i = CH_2OH$

Threonine (Thr) $R_i = \underset{|\;OH}{C}HCH_3$

Methionine (Met) $R_i = CH_2CH_2SCH_3$

Asparagine (Asn) $R_i = CH_2CONH_2$

Glutamine (Gln) $R_i = CH_2CH_2CONH_2$

Cysteine (Cys) $R_i = CH_2SH$

Acidic

Aspartic acid (Asp) $R_i = CH_2CO_2^-$

Glutamic acid (Glu) $R_i = CH_2CH_2CO_2^-$

Tyrosine (Tyr) $R_i = CH_2$– 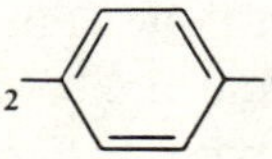 –OH (only weakly acidic)

Basic

Histidine (His) $R_i = CH_2$– 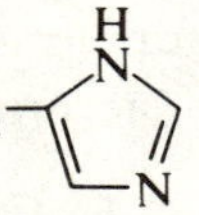

Lysine (Lys) $R_i = CH_2CH_2CH_2\overset{+}{N}H_3$

Arginine (Arg) $R_i = CH_2CH_2CH_2NH\underset{|\;NH_2}{C} = NH$

The resulting polymers are straight or globular chain molecules. The molecular masses of proteins (from the Greek for 'primary') vary between 10 000 and 1 000 000; the molecular masses of simple peptides are less than 10 000. As mentioned above, various side-chain residues R_i may be contained in these molecules. There are 20 different amino acids in living systems (see Table 3.1 for a list arranged according to neutral, basic and acidic residues). Owing to their steric structures free rotation about the C—N or C—C valences may be hindered. This effect greatly contributes to the formation of the secondary and tertiary structures of globular proteins, which are dealt with later.

Amino acids are 'zwitterions' (Structure (3.17)), a proton having been transferred from the carboxyl group to the amino group. In this case we speak of intramolecular salt formation. Amino acids are amphoteric. This is expressed by the following formulae:

$$H_3N^+CHRCOO^- + H^+ \rightarrow H_3N^+CHRCOOH$$
$$H_3N^+CHRCOO^- + OH^- \rightarrow H_2NCHRCOO^- + H_2O.$$

In the formation of the peptide bond, a covalent bond is formed between the α-amino group of one amino acid and the α-carboxyl group of another. In many cases heteropolar bonds (ionic bonds), hydrogen bonds, disulphide bonds (between the side chains of cysteine), dipole bonding forces, etc. also contribute to the formation of protein structure. Besides cellulose and nucleic acids the polymeric proteins are the most important constituents of all organisms.

As indicated above, the macromolecular structures of proteins fall into two broad groups, fibrous (insoluble) proteins and globular proteins.

(1) Fibrous proteins are characterized by a wide distribution of molecular weight but they have uniform structures along the chains. Therefore, they may well be compared to technical polymers. Their functions in the organism are easy to outline: formation of the supporting apparatus, the membranes, skin, ligaments and muscles.

Among the fibrous proteins are the keratins which form essential constituents of wool, hair, horn and claws. The keratins display helical structures of different lengths. Since only a few different amino acids are required for these simplest of proteins their structure and crystallinity can be investigated most easily. α-keratin (hair, wool and nails) consists of peptide chains arranged as right-handed helices. In hair there is an additional left-handed twist.

The collagens should also be mentioned. They are made up chiefly of glycine and proline.

(2) Globular proteins do not have such high molecular weights as fibrous proteins and are partly soluble. They are considerably more complicated than fibrous proteins as they form coils. Examples are myoglobin, haemoglobin, the hormone insulin and casein [1, 2, 3, 16, 44, 64, 73].

Besides the classification of proteins on account of their structure and chemical composition there are classifications which are based on their functions (Table 3.2).

Whilst discussing protein functions I should say something about enzymes. Enzymes, which act as biocatalysts, are particularly important to us. They act in starting, controlling, restraining and terminating chemical processes. The number of different macromolecular enzymes

Table 3.2. *Functions of proteins in living organisms*

Function	Characteristic protein(s)
(1) Formation of organ and cell structures (membranes, skins, supporting apparatus)	Keratin, collagen, membrane structure protein
(2) Hormones	Insulin, vasopressin
(3) Proteins with protective action	Antigens
(4) Toxins	Bacterial toxins
(5) Contractile proteins (muscles)	Myosin (actomyosin)
(6) Transport proteins	Haemoglobin (oxygen transport, blood), myoglobin (oxygen transport, muscle), serum albumin (fat and water transport)
(7) Storage proteins (no storage of information)	Casein (milk), albumin (white of egg)
(8) Organic catalysts (enzymes)	. . . ase

occurring in cells is estimated to be over 10 000. Their chemical designations are chosen depending on their biological functions, e.g., the term polymerase for an enzyme controlling the polymerization of DNA. To elucidate the structure of proteins certain chemical methods have been devised. Among these synthesis by the successive condensation of amino acids as well as degradation of proteins by successive removal of individual amino acids beginning at the end of the molecule deserve mention. Of the physical methods X-ray diffraction, mainly small-angle scattering, electron microscopy and the injection of radioisotopes should be mentioned. The unity of life allows us to apply the knowledge gained in microbiology to organisms made up of many differentiated cells [2, 3, 27, 64, 106*a*, 112*b*].

3.5 Synthesis of proteins by transcription and translation [44]

In the case of nucleic acids we have seen that template polymerization produces identical macromolecular nucleic acids. Here the decisive factor is the lock-and-key relation of the complementary purine and pyrimidine bases G and C, and A and T. Is there anything similar for proteins? One could imagine that the 20 amino acids form complementary pairs. Yet for steric reasons and on account of the different bonding character of the various amino acids this is not the case. Otherwise, each protein chain carrying different amino acid residues would have to play the role of a matrix with another one, similar to the former, being the patrix. This is impossible. Therefore, in forming identical proteins of a definite chemical composition nature makes a detour via RNA synthesis.

The following correlations obtain:

(1) All information and control instructions are fixed in DNA in material and individual-specific form.

(2) The structure and information content of DNA and the RNA resulting from it correspond in an identical manner to the proteins to be formed. This correlation is determined by the genetic code.

(3) There is a well-defined correlation between the primary structure of proteins, i.e., an exactly determined sequence of the amino acids R_i in the peptide chains, and the operations to be carried out by these molecules.

The sequence of the adenine, guanine, cytosine and thymine bases in the individual-specific macromolecular DNA molecules determines the amino acid sequence in the relevant protein molecules. This is called a sequence-conformal correlation. Thus, the structures of functional proteins which are typical of a particular living being are based on the polymeric DNA. The unique assignment of a protein to part of the DNA can be accomplished by coupling the amino acids to polyads in the sequence of nucleotides in DNA. This coupling principle is expressed in formula (3.9). This means that a particular sequence in DNA carries the information for the addition of a particular monomeric amino acid to the growing protein molecule.

Since there are only four nucleic acid bases but 20 amino acids unique assignment of one nucleic acid base to one amino acid is impossible. Also, the 16 possible diads of nucleotides are not sufficient to determine 20 different amino acids uniquely. (The 16 diads are: AA, AC, AG, AT, CA, CC, CG, CT, GA, GC, GG, GT, TA, TC, TG and TT.) Therefore, nature has to use a triad of nucleotides to determine one amino acid. As is easily seen, 64 different triads (codons) (AAA, AAC, etc.) can be formed in this way. These can be uniquely assigned to the 20 amino acids. The coupling of 20 amino acid residues to 64 different triads permits several variants for each amino acid. Hence, further instructions can be introduced into the code which is being established. These refer to chain initiation and termination and possibly to the recognition of errors in protein synthesis.

The relation between the DNA triads and the amino acids is called the genetic code. Its deciphering represents one of the most outstanding achievements of the twentieth century.

For steric reasons the sequence of the double-stranded DNA helix cannot be transferred to the protein to be formed. At first DNA must be transformed into a single-stranded form. This must take place without any change in the characteristic sequence of purine and pyrimidine bases present in the DNA and is done using RNA as the intermediate. (However, thymine is replaced by uracil in this process.) The single-stranded macromolecular RNA formed first is called messenger RNA (mRNA). The molecular weight of mRNA ranges to several million. Protein synthesis takes place in several steps:

(1) Single-stranded macromolecular mRNA is formed by copying from DNA. The former is sequence-conformal to the individual-specific DNA and hence a copy of the latter. The process is called transcription. It can be compared to the industrial printing process. The mRNA chain length ranges from 1000 to 15 000 monomeric units. mRNA has a lifetime of up to a few days and up to 100 protein molecules can be synthesized from each mRNA molecule by repeated translation. Recently, biological research has shown that besides informative series of nucleotides (genes) in DNA molecules there are ineffective parts in the chain molecules which probably mark the beginning or the end of a message, such as a TATA group. In transcription (i.e., during the formation of the single-stranded mRNA) these parts are eliminated.

(2) In the cytoplasm each of the 20 amino acids is carried by a specific transfer ribonucleic acid (tRNA). tRNAs have molecular weights up to about 2500, which means they are much smaller than mRNA. The synthesis of tRNA-amino acid complexes is brought about by specific enzymes (synthetases), with ATP

Fig. 3.2. Protein synthesis in a ribosome following transcription (schematic representation).

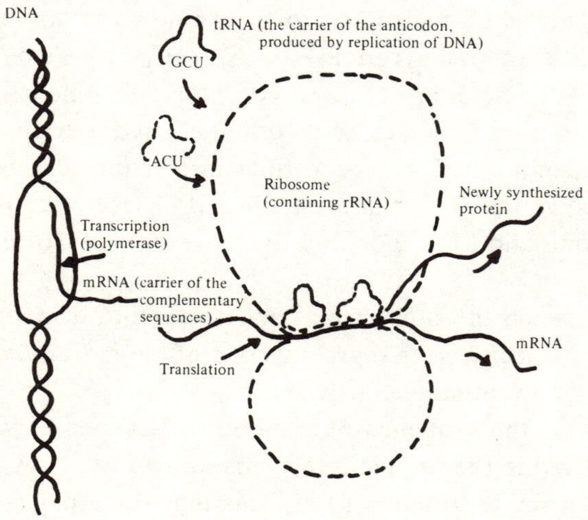

supplying energy. Also in tRNA uracil is present instead of thymine.

(3) With the participation of ribosomes tRNA complexes approach mRNA. Ribosomes consist of two ellipsoidal entities of different sizes (Fig. 3.2) in which protein synthesis takes place. Very generally speaking, this can be compared to the operation of a printing machine. A tRNA molecule is linked to the growing protein molecule only if three coordinated nucleotides are conformal to the respective sequence of nucleotides in the mRNA. In this way the correct sequence of the amino acids contained in the mRNA is secured. The sequence is free from overlap but till now no marks have been observed which make the triads recognizable. In translation polymerization the amino acids are linked to one another, but not to the nucleotides. After the short contact between these nucleotides including the amino acids they carry, the formation of the protein chain and the separation of the nucleotides from it take place. The basic nucleotide chain is absent in the separated protein which assumes a characteristic shape. Here the sequence of the chain polymer made up of 20 amino acid residues is decisive. The proteins should be considered as copolymers of 20 different monomeric units.

An extensive discussion of the control of the individual steps of protein synthesis to which magnesium ions also contribute is found in the fundamental work of Monod [52, 70], Crick [16] and Noll [74]. These authors introduced the terms 'operon', 'operator' and 'repressor' to characterize the control functions of the proteins initiating protein synthesis. These proteins control transcription and translation. Repressors have an inhibiting effect or cause synthesis to be terminated. The transcription and translation polymerization described in points (1) to (3) above is the most complicated but also the best controlled form of polymerization. Here 'translation' means a translation into the sequence of the proteins. In the genetic code all amino acids with the exception of methionine and tryptophan have several triads. In the different variants for each amino acid the first two bases are mostly unchanged. These variants differ mainly in the last base (wobble hypothesis).

Recent investigations [9, 14] have shown that a number of triads not occurring in the genetic code but appearing with the same probability in a random formation serve to transmit information about the termination of the reaction and to eliminate errors. The following statements can be made. For the initiation of protein synthesis, the codon AUG is responsible (in the presence of magnesium ions and ATP) and for

Table 3.3. *The genetic code*

Amino acid	Coordinated nucleotide triads
Alanine	GCA, GCC, GCG, GCU
Arginine	AGA, AGG, CGA, CGC, CGG, CGU
Aspartic acid	GAC, GAU
Asparagine	AAC, AAU
Cysteine	UGC, UGU
Glutamic acid	GAA, GAG
Glutamine	CAA, CAG
Glycine	GGA, GGC, GGG, GGU
Histidine	CAC, CAU
Isoleucine	AUA, AUC, AUU
Leucine	CUA, CUC, CUG, CUU, UUA, UUG
Lysine	AAA, AAG
Methionine	AUG
Phenylalanine	UUC, UUU
Proline	CCA, CCC, CCG, CCU
Serine	AGC, AGU, UCA, UCC, UCG, UCU
Threonine	ACA, ACC, ACG, ACU
Tryptophan	UGG
Tyrosine	UAC, UAU
Valine	GUA, GUC, GUG, GUU

termination the codons are UAA, UAG or UGA. Initiation, however, may be preprogrammed in RNA during transcription.

As stated above, the genetic code assigns the amino acids uniquely to the nucleotide triads [17]. An opposite assignment, such as the formation of nucleic acids controlled by amino acids is neither possible nor necessary. The genetic code is given in Table 3.3.

The following rules for protein synthesis have been established.

(1) The transfer of information takes place only in the direction DNA → RNA → protein.

(2) Each coding section of the DNA molecule, i.e., each gene, is associated with a specific protein. Each of these sections consists of a great number of nucleotides or amino acids [14].

Barrell, Air & Hutchinson [4] and Jung [54] showed that transcription may take place in the opposite direction, which means that deviations with respect to sequence (rule (1)) are possible. Further, it was found that overlap may occur, i.e., the coding regions for several different proteins may be assigned to one particular section of the DNA molecule if different starting places are permitted.

The formation of DNA and proteins demonstrates that the most important polymeric constituents of living matter are generated by self-reproduction and self-control. DNA and proteins constitute a whole, i.e., one common material basis for life processes equally present in all living beings. Yet the spatial variability due to changes in the sequence furnishes an enormous variety of chemically-different macromolecules.

4

Structure and structure formation in biomolecular systems

4.1 General principles of structure formation in biomolecules

The structures of single molecules and ensembles of molecules are determined by the laws of thermodynamics. This applies to biological systems as well without any restriction. In all cases the decisive requirement is thermodynamic equilibrium and this is determined by the extrema of free energy $F = U - TS$ or free enthalpy $G = H - TS$ (see Chapter 5). Often the equilibrium state that is valid for any temperature is established only very slowly or not at all in a finite time. Then macromolecular systems are frozen in a state of thermodynamic nonequilibrium. This applies to anisotropic molecular conformations, e.g., orientated molecular arrangements, and results in greater tensile strength in the directions of the molecular axes. The mechanical stability is further increased by helix formation and the twisting of several helices (double helix) [80]. Also, higher degrees of crystallinity give rise to increased mechanical stability. These general principles are realized in the formation of structure in cells, membranes, organisms, etc.

In some cases a particular structure may serve to fix some information. This is also utilized in technology. Thus, a definite location of a storage element in the matrix of a technical memory is associated with some definite information.

For biomolecules we distinguish:

(1) The structure of a single molecule, i.e., the relative positions (distances and bond angles) of atomic arrangements and the sequence of nucleotides or amino acids;

(2) the mutual coordination of neighbouring macromolecules, e.g., the formation of double helices (secondary structure);

(3) the formation of molecular coils, crystallites or clusters (tertiary structure);

(4) the coordination of several regions or crystallites to form higher-level units (quaternary structure).

In the field of synthetic polymers we have a quite similar classification of structures:

(1) The structure of a single molecule, for example in a solution;
(2) the combination of single molecules to form certain submicroscopic regions;
(3) the formation of crystals and partially crystalline regions;
(4) the appearance of supermolecular structures, such as spherulites.

For biomolecules we can state: a developing structure is correlated with its biological function. Thus, e.g., the increased stability resulting from crystal formation is utilized in the making up of cell walls, fibrils and membranes; fibres consist of orientated chainlike macromolecules; like rubberlike polymers, elastic elements and muscles have a large number of possible molecular conformations which are responsible for their elastic behaviour; the spatial alternation of side groups in syndiotactic polymers leads to a stronger tendency to crystallization. Nature makes use of the latter in the formation of cellulose where folded chains arise. This is shown schematically in Fig. 4.1.

Template polymerization determines the primary structure of biomolecules, i.e., the sequence of units in the complexly-structured molecules. There must be a correlation between the biological functions and the spatial structure of biomolecules. Nature strives to attain maximum storage capacity and a corresponding arrangement of the molecular units. However, such structuring of biomolecules cannot always be achieved with simple geometric forms.

The following principles should be borne in mind:

(1) The individual-specific DNA molecules must be structured in such a way that they contain all genetic information and all the different combinations of triads necessary for the synthesis of many thousands of different proteins. An additional requirement for the structure of some protein molecules is to take into account their character as catalysts (enzymes) for special complex chemical reactions.

Fig. 4.1. Graphic formula of cellulose illustrating syndiotactic organization.

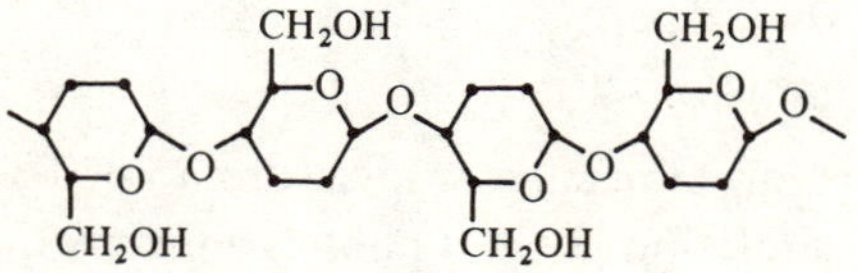

(2) To prevent molecule rupture by thermal vibrations molecules must be additionally stabilized by hydrogen bonds, donor-acceptor complexes, the formation of double helices and, in some cases, by sulphur bridges as well as by a variety of simultaneously-acting intermolecular forces.

(3) Intermolecular forces cause higher-level spatial structures, by giving rise to electrical and mechanical potentials and thus affecting biological operations.

(4) Through template polymerization the macromolecule acting as template transfers its degree of polymerization and sequence of units to the new molecule. Because the molecules are coupled to one another energetically the formation of higher-level structures is also transferred to the new system.

(5) The propagation rate is programmed so that most of the energy produced (exothermic process!) is given off to the environment. The process is isothermal rather than adiabatic. In addition, in certain cases the energy set free during polymerization is used to induce other and endothermic processes.

4.2 The primary structure of nucleic acids

The primary structure of nucleic acids is determined by the sequence of the bases along the chain molecules. Owing to the distinct information content of the genotypic storage every individual displays a particular sequence of nucleotides (P, phosphate groups; R, ribose):

```
P—R—P—R—P—R—.
  |    |    |
  A    G    G
```

The arrangement is asymmetric. There is a spiral twist from base to base. In general, DNA molecules do not form plane arrangements but form helices. The asymmetry causes optical rotatory properties. Another characteristic is that in all organisms the helices are wound in the same sense. Moreover, primary structure is determined by the specific content of adenine (A), thymine (T), guanine (G) and cytosine (C). Since in template polymerization A and T as well as G and C are mutually complementary, the contents of the complements involved are equal.

In many cases the ratio between the mole fractions of G and A is experimentally found to be 0.47. Then

$$G/A = 2G/2A = (G+C)/(A+T) \approx 0.47 \text{ [121].}$$

This is true for those parts of molecules which are joined together according to the lock-and-key relation. Recent publications say that there are loops of the structure where the lock-and-key relation is not realized and the ratio (G + C)/(A + T) may vary considerably.

The base sequence over certain parts of the very long DNA molecule is fixed in such a way that on the basis of the genetic code all the information required for the formation of a large number of proteins is also contained in it. Primary structure is individual-specific. It carries the genetic information and all the chemical information pertaining to the functional aspect. How these requirements can be met will be discussed in Chapters 6, 7 and 8. Every individual has its specific primary structure of DNA as distinct from that of all the other living beings that have lived or will live on earth. In replication this characteristic primary structure is duplicated and transferred to all new DNA molecules formed in the cells of that individual.

4.3 The primary structure of proteins [73, 97]

The nucleic acid–protein correlation on the basis of the genetic code implies regularities for the primary structure of the chain-forming protein molecules similar to those for nucleic acids. The number of amino acids in a polymeric protein molecule, however, is only one third of the number of nucleotides in the corresponding DNA.

The A chain in the double-stranded insulin specific to the cow starts with the sequence of amino acids

Gly—Ile—Val—Glu—Glu—Cys—Cys—Ala—Ser—Val—Cys etc.
(Glu: NH_2; Cys—S———S—Cys)

The sequences of amino acids in proteins are different for the various animal and plant species. This refers mainly to the end groups, e.g., for haemoglobin different structures are found in different animal species. Experiments have confirmed this also for immunoglobulins [56, 80]. Yet the middle parts of the protein chains where the catalytically active groups are located are more uniform. In addition, there are different hydrophilic and hydrophobic parts. These properties are of great importance for the secondary structure of proteins [73, 112*b*], which is dealt with below.

The primary structure of proteins was elucidated by the selective removal of end groups with complicated chemical methods in conjunction with analysis by means of chromatography and X-rays. The different chemical properties of amino acids facilitate these analytical methods and have made it possible to determine, e.g., the primary structure of haemoglobin. The genetic code allows conclusions to be drawn about the correlated nucleic acids. Conversely, it is hoped that the synthesis of particular proteins, such as that of insulin, can be carried out on a technical scale. It may be mentioned that recently this synthesis was accomplished with gene transfer using special microbes.

4.4 The secondary structure of nucleic acids [2, 20, 112*a*]

By secondary structure we understand the complementary arrangement of neighbouring DNA or RNA molecules and their winding into a double helix. Secondary structure arises from the interaction of van der Waals' forces and, in particular, of hydrogen bonds. These form between the oxygen and nitrogen atoms of the complementary pyrimidine and purine bases. The energetically most favourable structure is the DNA double helix (Fig. 4.2) which was first postulated and observed by Crick [15] and Watson [115, 116]. One helix turn comprises 9.3–11 base pairs depending on the crystal form (monoclinic or orthorhombic), is approximately 3.46 nm long and has a diameter of about 2 nm. The molecular planes of the purine and pyrimidine bases are almost normal to the helix axis and are directed inward. Since each time a purine and a pyrimidine base (characterized by a double ring and a single ring, respectively) complement each other the space between the strands is constant. Heating DNA to about 75 °C results in a progressive separation, i.e., denaturation, starting from the ends. The DNA is broken up into single strands. From the known energies of dipole bonding forces, hydrogen bonds and disper-

Fig. 4.2. DNA double helix after Crick & Watson.

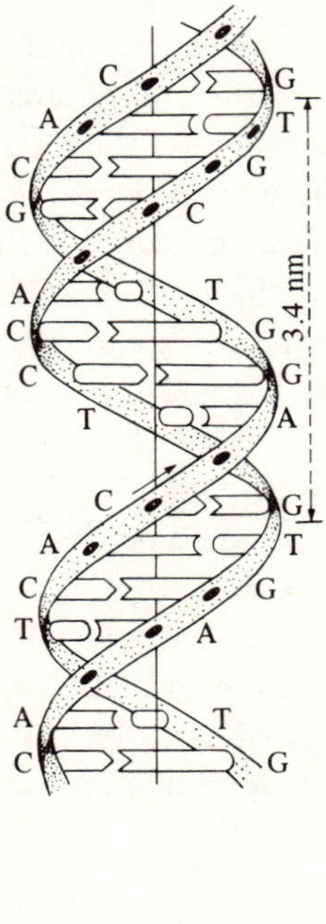

sion forces the energies of the pairs of opposite bases were calculated to be 6–8 kJ/mol [112*a*]. The linkage between G and C appears to be stronger than that between A and T.

In RNA uracil (U) is supposed to have a somewhat higher bonding energy than thymine. Hybrid structures consisting of DNA and RNA double strands are found. Sometimes they are folded or form double helices. In addition, the regular pairing of G and C and A and T may be interrupted by the false bonding of base pairs. This effect may lead to defects in the storage of information [20] or to the formation of segmental orders without any information content.

There are water molecules around the DNA double strand which compensate the electrical field effects caused by DNA. Recently zigzag structures which allow the double strand to be decoupled more readily have been under discussion. The double helix structure described above is characterized by optimum stability.

Unlike DNA, RNA does not display a helical structure but it also has complementary base pairs, CG and AU, whose coupling is mediated by hydrogen bridges. We find hairpin-like arrangements, which, in turn, form clover-leaf or starfish structures (Fig. 4.3). Finally, it should be noted that besides the well-known bases A, C, G, T and U, the rare bases 5-methyl cytosine [121] and 5-methyl oxycytosine occur in nucleic acids, giving rise to deviations in the secondary structure.

4.5 The secondary structure of proteins [97]

Some polypeptide chains (in globular proteins), consisting of 20 different amino acids in a well-defined sequence, are also wound right-handed about the molecular axis, with nearly the same angle of rotation from one amino acid to the next. In other cases they are folded (pleated

Fig. 4.3. Secondary structure of RNA after Doty [20].

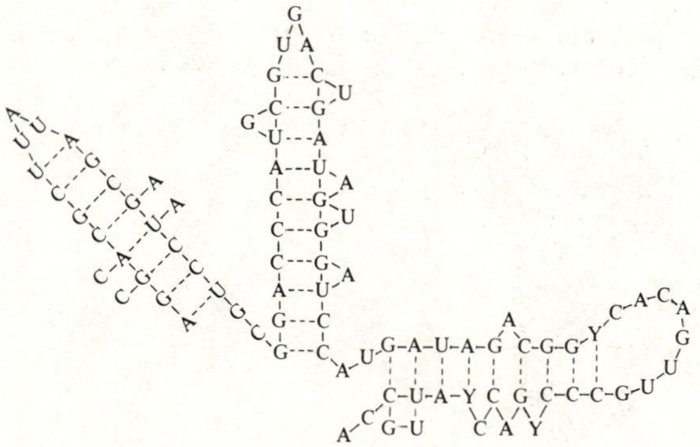

sheets). The observed structure is due to the asymmetry of the monomeric units. Helices are formed with the aid of hydrogen bridges and EDA complexes. The resulting molecules are relatively stable. In many cases the monomers are coupled by sulphide groups, intra- and intermolecular linkages. The latter have to be attributed to the primary valence bonds.

In fibrillar proteins (as in synthetic polymers, e.g., in nylon amides) there occur right-handed helices on which left-handed superhelices are superimposed. This phenomenon is called a coiled-coil structure. The fibrillar proteins present in wool, hair and epithelial cells display a strict periodicity. The helix investigated by Pauling & Corey [77] is characteristic of proteins. The observed results for various fibrillar proteins are different. Frequently 3.6 amino acid residues per turn are reported. After five turns (18 amino acid residues) the amino acids coincide with respect to a plane normal to the helix axis. The helix diameter was determined to be 1.01 nm not taking into account the amino acid residues directed outward. The height of each turn was found to be 0.56 nm. Hence the lattice constant in the direction of the helix axis is 2.8 nm. The different sizes of the side groups give rise to deviations of the helices from the ideal form. Considering intermolecular fields of force this must affect the formation of the tertiary and quaternary structures of proteins.

The effect of the secondary structure, i.e., the formation of an α-helix in the fibrillar protein is to increase its stability. While fibrillar proteins (such as α- and β-keratin and collagen) appear in the form of well-ordered helices globular proteins (single molecules) prefer a well-defined coil of molecular strings (Fig. 4.4) consisting partly of zigzag chains or short helices.

The interactions of the helix with the surrounding water are of importance. The water has a stabilizing effect on the secondary structure due to dipole bonding forces, hydrophobic forces within the coils and the association of water outside. In many cases noncross-linked proteins are

Fig. 4.4. Schematic representation of the tertiary structure of myoglobin [20] with the backbone chain drawn in.

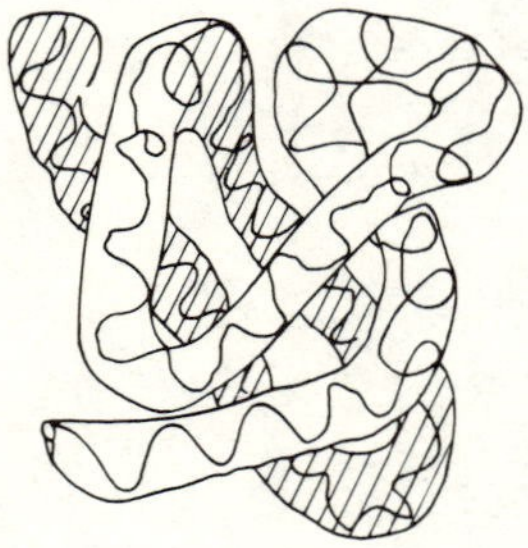

soluble in water. When heated the natural structure is destroyed and the biological efficacy is lost (denaturation). Finally, it should be pointed out that the protein chain helix is single stranded.

4.6 The tertiary and quaternary structures of proteins [2, 79]

The side groups in the protein chains are directed outward and together with strong dipole fields, hydrogen bonds and hydrophobic forces they cause steric folding of the molecular chains (Fig. 4.4). It has been found that in the formation of this tertiary structure the active groups relevant to the specific functions of enzymes are mostly directed towards the central regions of clusters. Tertiary structure is of great importance for the function to be fulfilled by an enzyme, e.g., to stop or hinder diffusion processes or the recombination of free radicals or excited states. A comparison with molecular sieves, such as zeolites, may serve to illustrate the effect of tertiary structure on the activity and kinetics of the processes. It must be said, however, that this comparison is incomplete.

Just as in nucleic acids the folding of double helices is characteristic of tertiary structure we find similar effects for the neighbouring parts in protein molecules. The most important enzymes, however, consist of several subunits. To repeat this point, the arrangement of polypeptide groups makes the protein backbone, with the atoms of the peptide linkage

Fig. 4.5. The structure of a cytochrome [2].

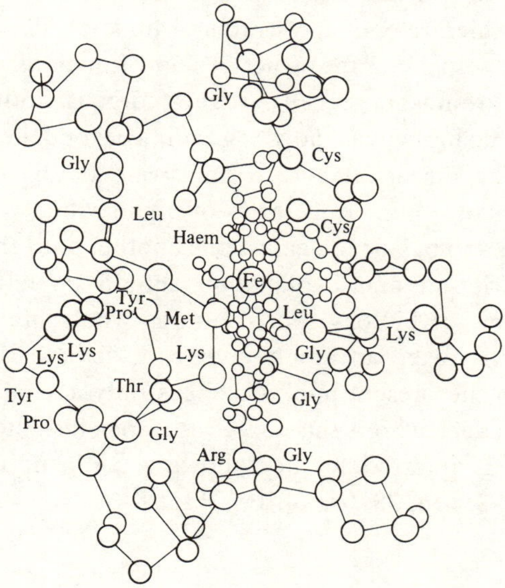

 lying in a plane. The rotation about the C—N bonds is hindered. The sequence fixed in the protein synthesis by the genetic code determines the secondary structure (zigzag chains or helices). The folding of neighbouring parts of the chain is called the tertiary structure and the spatial arrangement of subunits the quaternary structure. In globular proteins with molecular weights up to 100 000 no subunits are observed. This applies, e.g., to myoglobin, which is responsible for oxygen transport in the muscles. Haemoglobin, however, has four subunits, which are arranged in a tetrahedral quaternary structure.

Quaternary structure appears to be complex and difficult to trace. Nevertheless, just as is tertiary structure, it is of importance for the efficiency of proteins. Fig. 4.5 shows the well-known structure of cytochrome which is often presented in textbooks. It is one of the feats of modern structural chemistry to recognize such complex molecular structures and obtain unambiguous results, thus determining the locations of all atoms. To this, the X-ray diffraction technique, in particular small-angle scattering, has made a considerable contribution. For example, it was employed to measure the cube-like basic unit of myoglobin, giving the values of $4.6 \times 4.4 \times 2.5\ nm^3$ [73]. The latter protein consists of 153 amino acids.

The quaternary structures of many proteins are being studied by means of electron microscopy, in particular scanning electron microscopy, neutron scattering and X-ray diffraction investigations. Globules with diameters up to 100 nm have been found.

The steric structures of proteins and amino acids discussed above are the lowest level in the hierarchy of the structural units of life. It is not the aim of this book to elucidate the structures and functions of membranes, cell nuclei, chromosomes, cells, tissues, organs, individuals, species, etc. Therefore, no biological knowledge going beyond the concept of the molecule will be discussed here. I am aware of the fact that biological processes involve numerous levels of organization which are harmonically arranged and balanced against one another but I think that their exhaustive description and explanation should be left to the numerous compendiums of biology. For centuries whole branches of science have found this to be an ample sphere of action. Molecular biology as included in the present book describes only a small part of life as a whole. The reader who wants to get a deeper insight into the whole subject matter is referred to the following compendiums and reference books: [2, 3, 35, 56, 78, 87, 94, 102, 112*b*].

5

Energy and entropy in biomolecular systems

5.1 Fundamentals of thermodynamics

The laws of thermodynamics are valid for all biological processes without any restriction. There has never been, and will never be, any violation of the laws of nature, which are based on the principle of causality. Thermodynamics applies to all systems made up of a large number of molecules, atoms, electrons, radiation, etc. In physics, individual atoms or electrons are sometimes studied independently of their environment but mostly interactions between several bodies are investigated. Thus, a liquid may be considered together with its container, the vapour above it, the surrounding air, the support of the container, etc. All these bodies together constitute a thermodynamic system involving interactions. An initially hot liquid, for example, cools down to the temperature of its surroundings giving off heat to other parts of the system. If the surroundings are neglected the system is referred to as open. All living beings are in thermal contact with their environment. Taken separately, they represent open systems while together with their environment they form closed systems. In living beings chemical processes are continually taking place under isothermal and isobaric conditions. Unlike the technological processes for which we can choose temperatures and pressures as we please, in animate nature there are millions of chemical processes taking place under adverse conditions which are controlled by a huge number of catalysts, i.e., enzymes.

A brief account of the well-known laws of general thermodynamics may serve to give better insight into the thermodynamics of macromolecular systems. So let us consider thermodynamic states and the fundamental laws of thermodynamics.

What is a state? All bodies belonging to a system have a certain temperature, pressure, mass, density, volume, etc. These quantities are called variables of state. Some thermodynamic quantities are intensive (not proportional to mass), others are extensive (proportional to mass).

The description of a body involves more quantities than those given above, such as the quantity of heat. Part of this can be employed to do useful work, the efficiency depending on how the process is designed. Therefore, the quantity of heat is not a state variable and not necessary for determining the state of a system.

There are functional relationships between the state variables of the form $\phi(p, T, V) = 0$, where p is the pressure, T is the thermodynamic temperature and V is the volume. The relations between internal energy (U), temperature and volume or between enthalpy (H), pressure and temperature can be expressed by similar equations. They are readily derived for ideal gases in the form $pV = RT$ or, as the van der Waals equation $(p + a/V^2)(V - b) = RT$, where a and b are the constants of the van der Waals' law and R is the gas constant.

There are no fundamental thermodynamic equations known specifically for biological systems. In general, living beings are multiphase and consist of low-molecular weight aqueous components in which low- and macromolecular substances are dissolved, of organic liquids and mainly polymeric organic solids. Occasionally gas is enclosed.

5.2 The first law of thermodynamics

The first law of thermodynamics, which is obtained by applying the law of conservation of energy to thermodynamic systems, is of great importance. It reads

$$\Delta U = \Delta Q + \Delta A. \tag{5.1}$$

The work ΔA and the quantity of heat ΔQ are added to the system, thus changing its internal energy by ΔU. The first law of thermodynamics also describes energy transfers in chemical processes and these are taken to be positive in endothermic reactions (the system gains external energy) and negative in exothermic processes (the system loses energy).

In biological systems the transfer of energy is done by taking up quantum energy in photosynthesis as well as by supplying (giving off) chemical energy in transport processes, in producing entropy, in doing external work and in a variety of exothermic and endothermic processes which are controlled by the living beings themselves. Special mention should be made of the oxidation of iron (II) to iron (III), the oxidation of hydrogen sulphide, fermentation processes and the degradation of organic substances. A definite energy equivalent is associated with each chemical reaction in an organism. Some examples of energy changes

Table 5.1. *Energy sources for biological processes per mol*

	Energy transfer	
	kJ	kcal
(1) Photosynthesis (light absorption)		
$\lambda = 680$ nm (red light)	179	42.8
$\lambda = 400$ nm (violet light)	304	73.0
(The reduction of H_2O requires		
(water)	286	68.5
(water vapour),	242	57.9
the reduction of CO_2 requires (theoretically)·	474	114.5
Owing to repeated absorption of light quanta, however, the actual transfer amounts to	1540	369.0
This is an efficiency of 31%.)		
(2) Oxidation of H_2S to SO_2 and H_2O	276	65.6
(3) Oxidation of Fe^{II} to Fe^{III}	292	70.0
(4) Oxidation of $C_6H_{12}O_6$ to $6CO_2 + 6H_2O$ (of this, 40% is used for ATP formation from ADP; 60% is lost)	3708	886.0
(5) Fermentation to alcohol $C_6H_{12}O_6 \rightarrow 2C_2H_5OH + 2CO_2$	234	56.0
(6) Fermentation to lactic acid $C_6H_{12}O_6 \rightarrow 2CH_3CHOHCOOH$	196	47.0
(7) Conversion of adenosine triphosphate to adenosine diphosphate (ATP $\rightleftharpoons$ ADP)	29.2	7.0

associated with biological processes are given in Table 5.1. (Sometimes the electron volt is used as the unit of energy ($1\ eV = 1.6 \times 10^{19}$ J).)

While in plants quantum energy from the sun is the energy source, in animals it is the oxidative degradation of carbohydrates, fats and proteins that supplies the energy. In addition, living beings make use of numerous other chemical processes supplying energy, e.g., the oxidation of Fe(II) to Fe(III), that of hydrogen sulphide (H_2S) to sulphur dioxide (SO_2) and water (H_2O), and that of ammonia (NH_3) to nitrogen dioxide (NO_2) or nitrogen (N_2).

Energy transfer in organisms takes place in several steps:

(1) Energy gain by oxidative reactions;
(2) energy exchange by the transformation of adenosine diphosphate (ADP) into adenosine triphosphate (ATP);
(3) energy transport to the place where it is needed (influx of ATP);
(4) the conversion of ATP into ADP to enable a simultaneous endothermic process to take place.

 Enzymes are involved in all these processes, with ATP acting as an energy carrier. ADP and ATP are low-molecular weight compounds of adenine with ribose and two or three phosphoric acid residues respectively:

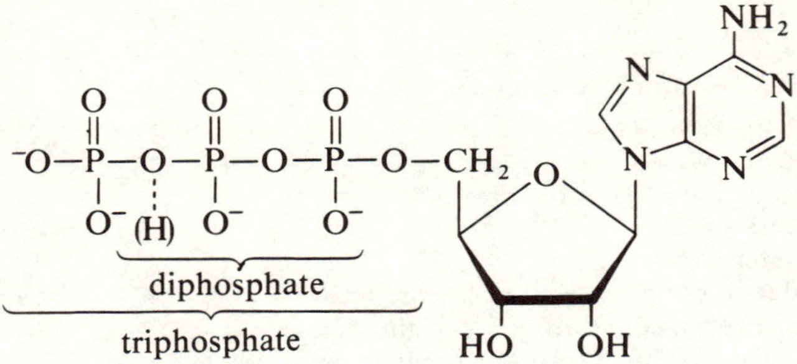

Structural formula of adenosine triphosphate

In triphosphate formation (phosphorylation) a proton from the phosphoric acid combines with the hydroxyl group of ADP to form water and an oxygen bridge is formed.

Many processes taking place in organisms are chain reactions in which the transfer of energy is governed by the first law of thermodynamics. In many cases exothermic and endothermic reactions are combined, the exothermic process continually using up the originating endothermic product and, on the other side, supplying energy to support the latter.

It seems necessary to illustrate assimilation and dissimilation as examples of the first law of thermodynamics in some detail [2, 56].

The decisive process for the energy balance of all living beings is the formation of organic substances, particularly starch, sugar and cellulose, in green plants. This is an autotrophic assimilation, i.e., a self-synthesis of organic substances from inorganic materials. It is a typical quantum-chemical process and the most cogent experimental evidence for quantum physics that can be found. In our teaching of physics we usually restrict ourselves to the photoeffect, the Compton effect, the laws of radiation and the interpretation of spectra in order to prove the fundamentals of quantum physics. Photosynthesis in green plants, however, shows us that the absorption of individual quanta of light with energies of about 179 kJ by one molecule provides enough energy to allow interchange processes and eventually the reduction of carbon dioxide to take place. For this purpose the quanta of light are absorbed in the active centre of a chlorophyll molecule. This gives rise to excited states. The surrounding parts of the molecule prevent recombination and contribute to storing energy delivered by absorption of several successive quanta of light. (The energy of a single quantum is not high enough to reduce one water molecule.) The first step is the oxidation of water. This takes place within the chloroplast and consists in forming $NADPH+H^+$ from $NADP^+$ (nicotinamide adenine dinucleotide phosphate) and H_2O. $NADP^+$ is the oxidizing agent of the first process (Hill process). Here also additional energy is present to generate ATP from ADP. The process can be represented by the following equation:

$$2H_2O + 2NADP^+ + n(ADP+P) + mh\nu \rightarrow O_2 + 2NADPH + 2H^+ + nATP$$

For the water we have the following equation:

$$H_2O \rightarrow \tfrac{1}{2}O_2 + 2H^+ + 2e^-.$$

The reaction taking place here is the photolysis of water with the release of two electrons and the formation of $NADPH+H^+$. The transfer of the electrons from a positive to a negative potential requires the energy coming from the absorption of light.

In a subsequent process (dark reaction) the reduction of carbon dioxide takes place with $NADPH+H^+$ as reducing agent and ATP as the supplier of energy. This can be represented by the following equation:

$$6CO_2 + 12H_2 \rightarrow C_6H_{12}O_6 + 6H_2O.$$

The resulting $NADP^+$ and ADP return to the starting process.

After this brief description of photosynthesis a second example of the first law of thermodynamics can be furnished by sugar oxidation, i.e., energy generation in the cell, on the basis of the Krebs cycle (citric acid cycle). This dissimilation is a process controlled by the cell nucleus and takes place in definite cell regions, the mitochondria, according to the principle of the division of labour. Like photosynthesis it is a basic process for all organic life and also proceeds in several stages. These are:

(1) The splitting of glucose into two molecules of pyruvic acid via several intermediate stages while two molecules of ADP take up energy and are converted to ATP.

(2) The combination of the pyruvic acid with the end product of the cycle to be described in (3), oxaloacetic acid, to form a citric acid molecule. This process involves some enzymes and passes through certain intermediate stages.

(3) The subsequent Krebs cycle proper in which eight different enzymes of highly complex structure participate in the formation of citric acid, aconitic acid, isocitric acid, oxalosuccinic acid, α-ketoglutaric acid, succinic acid, fumaric acid, malic acid and oxaloacetic acid.

In these stages the glucose molecule is broken down into water and carbon dioxide, a total of 38 molecules of energy-rich ATP being formed.

 The oxaloacetic acid starts a new cycle. In it, 1110 kJ instead of a possible 3708 kJ of energy are gained.

This brief description of assimilation and dissimilation is intended to show that the fundamentals of phenomenological thermodynamics can be applied to biological processes. It does not claim to supersede the detailed accounts of these processes given in the well-known textbooks and compendiums of biology.

5.3 The second law of thermodynamics, and reversible and irreversible processes

In inanimate nature, a system left to itself will be characterized by an irreversible increase in molecular disorder. The system left to itself tends to assume the most likely state.

In contrast to this, organisms follow a predetermined course of chemical reactions whose aim is to provide vital end products. This seems to be contradictory. In this section we shall have to prove that in living beings also there is a general tendency for the most likely state to be established. The irreversible processes responsible for this consist in equalizing temperature and pressure differences as well as concentration gradients and in damping mechanical motions. Thus, a directed molecular motion, such as a flow, eventually leads to disordered, direction-independent molecular vibrations and translations, i.e., to the generation of heat. Also, diffusion and mixing result in states of greater probability. This thermodynamic probability has to be discussed in more detail.

There may be many molecular arrangements representing a particular macroscopic state (involving many thousand millions of molecules). For definite variables of state the position and vibrational state of each single molecule may vary widely. For example, all molecules placed in the left-hand part of a vessel could vibrate quite strongly while those in the right-hand part are at rest.

In thermodynamics, the probability P is equal to the number of possibilities of establishing a particular (possible) macro-state. The interchange of the individual countable molecules results in new arrangements. If the distribution among the possible energy levels remains unchanged all arrangements belong to the same macro-state. Thermodynamic equilibrium is the macro-state having the greatest number of possible molecular arrangements. An unlikely state may occur now and then but it will be the rarer the more it deviates from thermodynamic equilibrium.

Natural processes proceed in the direction of growing disorder. In molecular terms this means equalizing the concentrations and the levels of thermal energies in the individual volume elements and hence mixing and the elimination of temperature and pressure differences. Regular arrangements, e.g., $a-b-a-b-a-b$ etc., or $a-a-a-a-a-b-b-b-b-b$ are replaced by irregular ones such as $a-a-b-a-b-b-b-a-a-b-a-b$ etc., where a and b may be molecules of two different kinds or molecules vibrating more or less strongly, respectively. A state of greater thermodynamic probability will never change by itself into a less likely one. In nature the law

$$P_{\text{small}} \rightarrow P_{\text{great}}$$

is obeyed.

Physicists are not satisfied with this qualitative statement. They look for a measurable quantity assigned to the system. This quantity should be a measure for the probability of a certain state. It should vary with changes of state but assume its initial value if in a reversible process the original state is reached again. After irreversible processes it will not be possible to measure again the same value for this characteristic quantity describing the system without extending the system. The change of a thermodynamic system from an old to a new state gives different values for this quantity. But this must not depend on the kind of transition to the new state. If an introduced variable S depends on other quantities, say x and y, the change $\mathrm{d}S$ will be a unique function of x and y only if

$$\mathrm{d}S=\frac{\partial S}{\partial x}\mathrm{d}x+\frac{\partial S}{\partial y}\mathrm{d}y \quad \text{and} \quad \frac{\partial^2 S}{\partial x\,\partial y}=\frac{\partial^2 S}{\partial y\,\partial x}.$$

With regard to the first law of thermodynamics

$$\mathrm{d}Q=\mathrm{d}U+p\,\mathrm{d}V \quad \text{with } p\,\mathrm{d}V=-\mathrm{d}A \text{ according to (5.1).}$$

This requirement is satisfied only if we divide by the thermodynamic temperature T. We set $\mathrm{d}U=c_v\,\mathrm{d}T$ and get

$$\mathrm{d}S=\frac{\mathrm{d}Q}{T}=c_v\frac{\mathrm{d}T}{T}+\frac{p\,\mathrm{d}V}{T} \tag{5.2}$$

where Q is the quantity of heat, U is the internal energy, p is the pressure, V is the volume and c_v is the specific heat at constant volume. $\mathrm{d}S=\mathrm{d}Q/T$ is a total differential. With $p=RT/V$ (equation of state for an ideal gas) we obtain

$$\Delta S=c_v\ln\frac{T_2}{T_1}+R\ln\frac{V_2}{V_1}. \tag{5.3}$$

The change ΔS is independent of the path. Therefore, S is a state variable which, together with other variables, is very suitable for describing the instantaneous state of the system. The variable which S represents is called entropy. We have yet to prove that the state variable S thus defined, or its change $\Delta S = \Delta Q/T$ is related to the probability of finding the system in a definite state. If we allow the system to change at constant temperature ($T_1 = T_2$), then

$$\Delta S = \dot{R} \ln \frac{V_2}{V_1} \tag{5.4}$$

(isothermal change of state of an ideal gas).

A similar calculation applies to the mixing of several substances, gases or liquids, the temperature remaining unchanged. Integration of eqn (5.2) yields the entropy as a state variable to which an arbitrary integration constant may be added. In general, for any substance we have

$$\mathrm{d}S = \frac{1}{T}\left(\mathrm{d}U - \sum_a X_a \,\mathrm{d}x_a - \sum_i \mu_i \,\mathrm{d}N_i\right) \tag{5.5}$$

where N_i is the number of particles (mole number), μ_i is the chemical potential of the ith component and X_a represents generalized forces such as Δp_a (pressure difference), ΔT_a (temperature difference), and Δc_a (concentration difference). X_a is given by the thermal equation of state $X_a(T, V, p)$ or the caloric equation of state $X_a(T, U, V)$.

In order to establish a numerical value for the entropy, say, of water vapour in technology it has been decided to take the integral

$$\Delta S = \int_1^2 \frac{\mathrm{d}Q}{T} = \int_{T_1}^{T_2} \frac{c_p}{T} \mathrm{d}T \tag{5.6}$$

(where c_p is the specific heat at constant pressure) from the lower limit $T_1 = 273.15$ K to T_2.

In mixing processes, in the flow of gases into vessels and in temperature equalization by heat conduction and radiation the quantity $\Delta S = \int_1^2 \mathrm{d}Q/T$ will increase. Nothing happens in animate and inanimate nature where such an irreversible process can be reversed without an external influence, i.e., an extension of the system. This is what the second law of thermodynamics states. It is valid for closed systems.

If in the first law of thermodynamics

$$\Delta U = \Delta Q + \Delta A$$

we denote the obtainable work ΔA by ΔF (free energy), division by T gives (for isothermal processes):

$$\frac{\Delta F}{T} = \frac{\Delta U}{T} - \frac{\Delta Q}{T} \quad \text{or with} \quad \frac{\Delta Q}{T} = \Delta S$$

$$\Delta F = \Delta U - \Delta S \cdot T \quad \text{or} \quad \mathrm{d}F = \mathrm{d}(U - ST) = 0 \tag{5.7}$$

in equilibrium.

Setting $\Delta F + p\,\Delta V = \Delta G$ and $\Delta U + p\,\Delta V = \Delta H$ we get

$$\Delta G = \Delta H - \Delta S\,T \text{ or}$$

$$dG = d(H - ST) = 0 \tag{5.8}$$

in equilibrium where H is the enthalpy. Only the work ΔF is available as free energy or ΔG as free enthalpy. A body left to itself will irreversibly change until its free energy or free enthalpy has reached a minimum. This is done by increasing ΔS. In chemical changes taking place at constant temperature and pressure, ΔH is negative for exothermic reactions whereas it is positive for endothermic ones. This means that in the latter reactions energy must be supplied to the system from outside.

The thermodynamic probability P can be connected with the entropy defined by the relation $dS = dQ/T$ (5.2). This is accomplished using the formula

$$S = \text{const.} \cdot \ln P. \tag{5.9}$$

Let us derive this for a simple case. Let us consider two different volumes of an ideal gas, V_1 and V_2, V_2 being the larger. We ask how many times greater the probability of finding a particular number of molecules in the total volume V_2 will be than that of finding them together in the smaller volume V_1 at a given instant. The result for one molecule will be the ratio V_2/V_1. For two molecules (A and B) the ratio will be $(V_2/V_1)^2$. For N_A molecules, i.e., for 1 mol of gas, the probability of finding them in the total volume V_2 is enormously much greater than that of finding them in the partial volume V_1. In practice the gas will fill the whole space V_2 and the number of molecular arrangements increases $(V_2/V_1)^{N_A}$ times. Thus the increase in thermodynamic probability is

$$\Delta P = \left(\frac{V_2}{V_1}\right)^{N_A}.$$

Taking the logarithm we get

$$\ln \Delta P = N_A \ln \frac{V_2}{V_1}. \tag{5.10}$$

This equation reminds us of the relation (5.4) which is valid for isothermal changes of state after introducing the integrating factor $1/T$ into the first law of thermodynamics:

$$\Delta S = R \ln \frac{V_2}{V_1} \quad \text{or} \quad \Delta S = R \ln \frac{c_1}{c_2} \quad \text{with } c_1 < c_2$$

where c is the concentration of the gas.

The calculated rise in thermodynamic probability with increasing volume from V_1 to V_2 corresponds to the isothermal expansion described by eqn (5.4). Thus, on eliminating $\ln V_2/V_1$ from eqns (5.4) and (5.10) we have

$$\Delta S = \frac{R}{N_A} \ln \Delta P.$$

With $k = R/N_A$ this gives

$$\Delta S = k \ln \Delta P \qquad (5.11)$$

where k is the well-known Boltzmann constant (1.38×10^{-23} J/K). The increase in entropy is proportional to the increase in the ratio between likely and unlikely cases. In the above example it is proportional to the logarithm of the volume ratio.

By definition the probability considerations put forward here for an ideal gas also apply to real gases, liquids and solids. The transition from less likely to more likely arrangements, i.e., the irreversible equalization of temperatures and pressures, mixing and the conversion of ordered to disordered molecular motion increases thermodynamic probability and hence entropy. For all irreversible processes we have

$$\Delta S_{irr} > 0. \qquad (5.12)$$

Establishing that entropy is proportional to the logarithm of thermodynamic probability was Ludwig Boltzmann's greatest achievement. Entropy is one of the most important state variables. Its significance goes beyond the integral

$$\Delta S = \int_1^2 \frac{dQ}{T}$$

which is valid for reversible processes (see (5.6)). If repeated measurement of the entropy of a closed system yields different values S_1 and S_2 with $S_2 > S_1$ the entropy difference $S_2 - S_1 = \Delta S > 0$ is due to irreversible processes that have taken place between the measurements. On account of the irreversibility the value $S_2 > S_1$ must have been taken later than S_1.

A system which is not in thermal equilibrium will continue to change by itself until the entropy S has reached its maximum and the free energy $F = U - TS$ its minimum. Then thermodynamic equilibrium is established. If the molecules in a body are free to move thermodynamic equilibrium is reached very rapidly.

5.4 The second law of thermodynamics in macromolecular, particularly biological, systems

In macromolecular systems we can frequently observe apparent deviations from thermodynamic equilibrium. There are various reasons for this:

(1) *The freezing-in of high-temperature equilibrium states during cooling down.* Owing to the thermodynamic potentials $G = H - TS$ or $F = U - TS$ a definite amount of entropy S is associated with each temperature when a system is in equilibrium. Cooling results in molecular dislocations leading to the formation of ordered structures owing to the forces acting between molecules. The transition takes place over potential barriers which must be overcome if one molecular conformation changes from one probable position to an adjacent one. (This motion is hindered, e.g., by bond angle distortions.) Fig. 5.1 shows a projection of chemical potentials caused by bonding forces with two positions, 1 and 2, separated by the activation energies ΔU_i and ΔU_k. In order to establish thermodynamic equilibrium the potential barriers must be overcome by thermal vibrations [48]. During rapid cooling the probability for thermal vibrations with amplitudes ΔU_i and ΔU_k is too low and the molecular conformations remain unchanged in the position which they had in the high-temperature region. Thus we have frozen the body in a state of higher entropy.

A similar case is the nonestablishment of a chemical equilibrium which would be expected according to the law of mass action. Thus, detonating gas may be stable for any period of time. The activation energy necessary to establish equilibrium in an explosion is provided only by thermal excitation, i.e., ignition of the gas mixture. The existing potential barriers, or activation energies can, however, be lowered by means of catalysts. This drastically speeds up the establishment of chemical equilibrium.

Certain molecular conformations, such as orientations, can be imposed by means of external forces and fields. The structures thus characterized

Fig. 5.1. Section of an energy hyperplane between two possible macromolecular conformations. Vertical hatching, occupation of possible positions at high temperatures; oblique hatching, the same at low temperatures; 1, 2, positions of chemical potentials; ΔU_i, ΔU_k, activation energies.

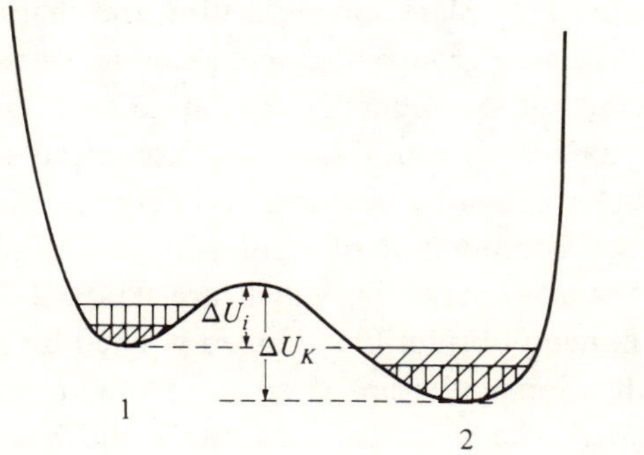

 by diminished entropy remain unchanged on cooling. Heating leads to the structure and shape corresponding to thermodynamic equilibrium. A stretched plastic body contracts on heating. At sufficiently low temperatures the external mechanical stress can be removed without changing the shape or structure and without reaching thermodynamic equilibrium. Note again that at low temperatures the thermal molecular motion is not strong enough to make available the activation energy ΔU required to establish equilibrium in a sufficiently short time.

The rate of change $\mathrm{d}S_n/\mathrm{d}t$ of a thermodynamic nonequilibrium characterized by the entropy S_n is frequently proportional to the difference $S_n - S_e$ where S_n is the entropy at a given time t, S_0 is the entropy for $t = 0$ and S_e is the equilibrium entropy. The rate constant is denoted by $K(T)$. We have

$$\frac{\mathrm{d}S_n}{\mathrm{d}t} = -K(S_n - S_e). \tag{5.13}$$

The solution is

$$S_n = S_e - (S_e - S_0)\,\mathrm{e}^{-Kt}. \tag{5.14}$$

After Eigen [25] the rate $\mathrm{d}S/\mathrm{d}t$ is proportional to the product of forces and fluxes. It is evident that generalized forces may be interpreted in terms of affinities A between partners of a reaction. So the rate constant $K(T)$ for the establishment of equilibrium becomes proportional to A. $1/K$ is a relaxation time that depends exponentially on an activation energy.

(2) *The fixation of definite molecular arrangements by means of template polymerization.* Frozen-in nonequilibria are typical of macromolecular systems. This is true not only with respect to the positions of different molecules relative to one another but also (in particular) for the single molecule as a whole. Thus the formation of very special configurations is possible. Owing to their structures these are systems far from thermodynamic equilibrium.

Macromolecules formed by template polymerization and characterized by definite sequences of the side groups adenine, guanine, cytosine and thymine (DNA) or by exactly fixed sequences of amino acids (proteins) represent such thermodynamic nonequilibria. A macromolecule bearing a certain preferred sequence would be most unlikely to be formed randomly or by a stochastic combination of monomers. This applies also to the secondary and tertiary structures, which are coupled with the primary structure. The nonequilibrium first formed is fixed by template polymerization (by replication and translation) and identically transferred to further molecules. The sequences are stable and have only a very small rate of change towards stochastic sequences. This feature of macromolecular systems has received little attention so far. Information coupled with the frozen-in structure may be fixed for any length of time. Thus macromolecules are stable storehouses of information over a rather wide temperature range and may remain in this thermodynamic non-equilibrium. The transition to another sequence requires the overcoming of a certain energy barrier ΔU which is of the order of 100–400 kJ if covalent bonds are involved. The chemical bonds characterized by such high activation energies are stable from 0 K to about 500 K.

As regards biological and chemical processes, energy barriers ΔU_i (activation energies) preventing the establishment of thermodynamic equilibrium are diminished by catalysts (enzymes). These tend to speed up the equilibration process. The converse is also possible.

5.5 The conservation of thermodynamic states far from thermodynamic equilibrium by cybernetic processes

Nonequilibria can be maintained as long as one pleases with the aid of external forces or free energy added to the system from outside acting against the irreversible alteration. An everyday example is the conservation of the unstable state in juggling. Countermovements are made to prevent any change of state away from the unstable equilibrium. The same applies to cycling and rope-dancing. According to Taylor's law, in the first time interval Δt a thermodynamic nonequilibrium S_{n} changes by

$$\Delta S_{\mathrm{n}} = \frac{\mathrm{d}S_{\mathrm{n}}}{\mathrm{d}t}\Delta t, \qquad \Delta S_{\mathrm{n}} > 0. \tag{5.15}$$

The free energy $\Delta F = -(\mathrm{d}F/\mathrm{d}t)\Delta t$ has to be fed into the system from outside to counterbalance this entropy change (for systems with constant internal energy). Then,

$$\mathrm{d}F = \mathrm{d}S_{\mathrm{n}}T. \tag{5.16}$$

The system produces the entropy $\mathrm{d}S_{\mathrm{n}}$ which is compensated by free energy. On integration we get

$$\Delta F = \int_0^{\Delta t} \frac{\mathrm{d}S_{\mathrm{n}}}{\mathrm{d}t} T\,\mathrm{d}t > 0. \tag{5.17}$$

ΔF is small if the disturbance is compensated very rapidly (Δt small). An experienced cyclist will restore equilibrium more quickly than a learner, that is, he will need to use less energy for steering. The conservation of undamped oscillations by a stimulated energy supply may also be mentioned here. In a clock the energy of the pendulum lost due to friction is supplied anew in phase by the spring. In a laser the energy given off is replaced by the atomic rearrangement stimulated in phase by the light itself. Thus a laser is an analogue of the feedback which is of decisive importance in organisms.

Dissipative structures (flow equilibria) are systems far from thermal equilibrium in which entropy is continually produced, in general without changing the internal energy. They remain in the state of thermal non-equilibrium. Flow equilibria are characterized by generalized forces (e.g., affinities) and related fluxes (chemical reactions) [23].

The change $\mathrm{d}Z$ of an extensive quantity Z is divided into a part that is subject to exchange (subscript e) and a part that is inherent in the system (subscript i). This leads to the relation

$$\mathrm{d}Z = \mathrm{d}Z_\mathrm{i} + \mathrm{d}Z_\mathrm{e}.$$

If we assume our extensive quantity to be the entropy S, then, according to Prigogine [84],

$$\mathrm{d}S = \mathrm{d}S_\mathrm{i} + \mathrm{d}S_\mathrm{e}. \tag{5.18}$$

Following Prigogine we use symbols without subscripts for quantities of the total system; the subscripts L and D are assigned to living and dead systems respectively.

If an open system under consideration is characterized by giving off more entropy ($\mathrm{d}S_\mathrm{e} < 0$) than is produced in the interior ($\mathrm{d}S_\mathrm{i} > 0$) the entropy diminishes:

$$\mathrm{d}S = \mathrm{d}S_\mathrm{i} + \mathrm{d}S_\mathrm{e} < 0, \qquad |\mathrm{d}S_\mathrm{e}| > |\mathrm{d}S_\mathrm{i}|. \tag{5.19}$$

In the case of equilibrium, $\mathrm{d}S_\mathrm{i} = -\mathrm{d}S_\mathrm{e}$, and hence $\mathrm{d}S = 0$. Dissipative systems can form only when there is a critical distance from thermodynamic equilibrium, i.e., a thermodynamic nonequilibrium.

Dissipative structures far from equilibrium are described by nonlinear differential equations and nonlinear fluxes [36, 38, 84]. The production of internal entropy can be presented as the action of fluxes I_j and forces X_j:

$$\frac{\mathrm{d}S_\mathrm{i}}{\mathrm{d}t} = \sum_j I_j X_j \geqslant 0.$$

According to Eigen [25] and Prigogine [84] dissipative systems manifest themselves by autocatalytic processes. Prigogine proves that in any dissipative system the production of entropy is delayed until for flow equilibrium the value $\mathrm{d}S_\mathrm{i}/\mathrm{d}t = 0$ is reached. From the equilibrium conditions

$$S_\mathrm{i} = \mathrm{max}! \quad \text{and} \quad \frac{\mathrm{d}S_\mathrm{i}}{\mathrm{d}t} \geqslant 0$$

we get

$$\frac{d^2S_i}{dt^2} \leq 0 \tag{5.20}$$

('physical evolution').

In a dissipative system the forces (affinities) cause the rate of entropy production to decrease. This relation also applies to the case of nonconstant internal energy ($dU/dt \neq 0$). The existence of a flow equilibrium is causally related to oscillations or fluctuations about the nonequilibrium state [84, 85].

According to Bertalanffy [6, 7], for the steady state we have

$$dU_e = 0, \qquad dS_e = -dS_i \leq 0.$$

Negentropy must be supplied to the system to compensate the entropy production. I introduce the term negentropy for negative entropy with the purpose of avoiding negative numbers. As entropy increases negentropy decreases. In the thermodynamic description steady-state systems are open and nonlinear. In his fundamental work Ebeling [23] has calculated the nonlinear relations for the stability of such oscillating systems. The oscillations appear suddenly and increase in amplitude until a steady state (flow equilibrium) is reached like the growth of oscillations in a radiofrequency generator or the transition from laminar to turbulent flow in hydrodynamics. In every case a certain experimental set-up (corresponding to some basic information) or an unstable physical or chemical state is required. Natural systems exhibit a damping of the oscillations. Dissipative systems display fluctuations of entropy. This is true, e.g., for the ADP $\leftrightarrows$ ATP conversion [10].

The theory of dissipative open systems describes dynamic processes. Therefore, in all cases the rate of thermodynamic change is of interest.

Fig. 5.2. Block diagram of a flow equilibrium maintained by a feedback reaction.

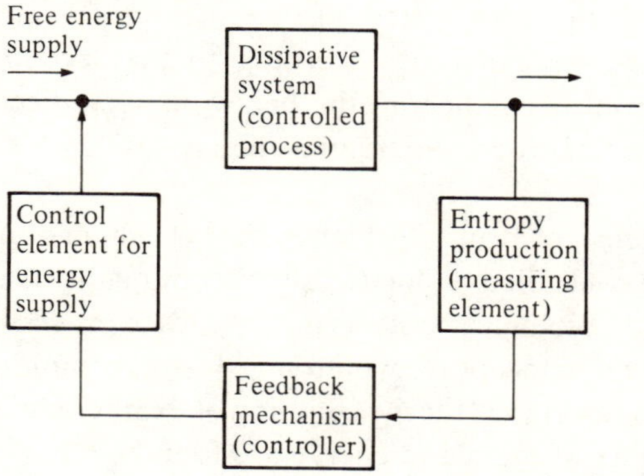

 In treating dissipative systems the rate of entropy production is as important as the consideration of entropy itself. The relevant nonlinear differential equations may have several solutions, the relation $dS_i = 0$ being valid for maintaining the flow equilibrium or dissipative state far from thermodynamic equilibrium.

The fact that dissipative systems can be formed from nonequilibria makes it possible to accomplish flow equilibria characterized by the input of free energy and maintained by a feedback reaction (Fig. 5.2). This possibility, which will be discussed in Section 11.1, extends the scope of the cases treated in [23].

The generation of dissipative systems may be considered as a kinetic phase transition. The living being as an open thermodynamic system is far from thermodynamic equilibrium and possesses entropy, S_n. If it were left to itself (i.e. without feedback or autocatalysis) any nonfrozen-in unstable state would approach thermodynamic equilibrium at a rate of $dS_n/dt > 0$. In dissipative systems the deviation from equilibrium is determined by receptors, or, in the case of living beings, by enzymes acting as 'sensors'. They control the supply of free energy to the open system thus maintaining the latter. This can be illustrated by the following scheme:

Thermodynamic nonequilibrium dissipative system		Thermodynamic equilibrium	
S_n	$\rightarrow$	$\left(\frac{dS}{dt} = 0\right)$	$S \rightarrow$
(oscillations about S_n)			

There is a delay, sometimes inevitable, in the supply of free energy. At the same time a sign reversal is brought about in the sense that the free energy introduced into the open system just counteracts the increase in entropy production to be expected according to the second law of thermodynamics. In control engineering this behaviour is studied in detail and described by nth order differential equations with constant coefficients.

Cases of programmed control in which a desired change of the set (desired) value of the controlled biological quantity or quality takes place are also investigated. In biological systems this applies to processes which are characterized by the concept of evolution. Mostly, complex control circuits which are of importance for the coupling of exothermic reactions with a controlled influx of mass and energy are found.

For constant internal energy ($\mathrm{d}U=0$), it follows from the basic law
($\mathrm{d}F=\mathrm{d}U-T\,\mathrm{d}S$) that for isothermal processes with $\mathrm{d}S=\mathrm{d}S_\mathrm{i}+\mathrm{d}S_\mathrm{e}$:

$$\frac{\mathrm{d}S_\mathrm{i}}{\mathrm{d}t}=-\frac{\mathrm{d}S_\mathrm{e}}{\mathrm{d}t}-\frac{1}{T}\frac{\mathrm{d}F}{\mathrm{d}t} \tag{5.21}$$

where

$$\frac{\mathrm{d}S_\mathrm{i}}{\mathrm{d}t}=0 \quad \text{for} \frac{\mathrm{d}S_\mathrm{e}}{\mathrm{d}t}=-\frac{1}{T}\frac{\mathrm{d}F}{\mathrm{d}t}.$$

The decrease in entropy caused by the evolution, growth and increase of information of living systems must be compensated from outside by supplying free energy $\mathrm{d}F_\mathrm{e}$. With constant internal energy we have

$$\frac{\mathrm{d}F_\mathrm{e}}{T}=-\mathrm{d}S_\mathrm{e}=\mathrm{d}S_\mathrm{i}-\mathrm{d}S.$$

Ageing, the decay of bodies and the death of cells and an individual are described in thermodynamic terms by

$$\frac{\mathrm{d}S}{\mathrm{d}t}>0.$$

The free energy supplied to maintain the flow equilibrium, i.e., the dissipative structure of the system, is determined by the abundance of the external energy source available, e.g., the intensity of the quantum radiation in the case of photosynthesis, or by the energy flux demanded by the control unit (feedback). The thermodynamic equation (5.21) may serve to describe a control mechanism. The energy supplied and the entropy changes balanced by the former have opposite signs.

If we consider changes in internal energy ΔU of the system we have

$$\Delta U=\sum_j F_j+\sum_k F_k+T\sum_m \Delta S_m \tag{5.22}$$

where $\Delta U<0$ and $\Delta U>0$ for exothermic and endothermic processes respectively, and

- ΔU is the change in internal energy of the system (e.g., energy storage owing to the uptake of food, formation of ATP from ADP, etc., growth);
- ΔF_j is the free energy introduced into the system (e.g., quantum radiation, ATP formed, the combustion of glucose according to the Krebs cycle, or energy influx to maintain thermodynamic nonequilibrium, i.e., to control the flow equilibrium) ($\Delta F_j>0$);
- ΔF_k is the work that may have to be done (e.g., muscular work, diffusion against the natural concentration gradient) ($\Delta F_k<0$);
- ΔS_m is the entropy change (e.g., the rate of flow of heat to the environment, passive diffusion, heat production by internal friction).

 Cell degradation and ageing require positive contributions to the system; the formation of more complex, i.e., more informative molecular arrangements in evolution and information storage imply negative contributions.

Living systems, which are far from thermodynamic equilibrium, remain in this nonequilibrium. Energy transformations are steadily taking place in them. They use the nonequilibria existing in nature, such as quantum radiation or the possible oxidation of inorganic and organic substances, as 'high-grade' external energy to be exchanged, and give off 'low-grade' energy (heat, work) to the environment. At the same time they make provision for the entropy production taking place within them by taking up free energy. In general, for the open system of a living organism with entropy S_L,

$$\frac{dS_L}{dt} \leqslant 0. \tag{5.23}$$

Of course, we should mention cases in which cell degeneration occurs, e.g., in the formation of cancerous tissue. Here, $dS_L/dt > 0$.

If, however, we consider a living system together with its environment (entropy S_{en}), we have

$$\frac{dS_L}{dt} + \frac{dS_{en}}{dt} \geqslant 0 \tag{5.24}$$

where

$$\left|\frac{dS_{en}}{dt}\right| \geqslant \left|\frac{dS_L}{dt}\right|. \tag{5.25}$$

In the case of death (entropy S_D) all control processes are stopped, and then, for the originally open system,

$$\frac{dS_D}{dt} \geqslant 0. \tag{5.26}$$

As an extension of Schrödinger's [95] fundamental considerations we can state that living beings can change their contact with the environment either consciously or unconsciously but always necessarily and appropriately. This influences the flux of entropy and results in the control of free energy consumption [46]. Eqns (5.15)–(5.20) are necessary additions to the laws of thermodynamics referring to biological processes. They are part of the framework of the axioms of thermodynamics. Organisms are not, of course, allowed to utilize even obvious deviations from regular thermodynamic behaviour as energy sources. Biological systems cannot select from an ensemble of molecules characterized by a broad distribution of thermal vibrational energy (Boltzmann distribution of energy) the more energetic ones to produce energy for themselves; Maxwell's 'demon' cannot be accomplished, life cannot circumvent the laws of nature in any point. Furthermore, the second law permits only a maximum efficiency determined by the originally given temperature intervals. The maximum efficiency is

$$(T_1 - T_2)/T_1 < 1 \tag{5.27}$$

where T_1 is the temperature of the upper energy reservoir and T_2 is that of the lower one. T_2 is mostly assumed to be ambient temperature.

In isothermal processes the temperature intervals have to be replaced by chemical affinities. These represent the generalized 'forces' for the dynamics of thermodynamic transformations. The primary utilization of solar quantum radiation in plants gives rise to a high concentration of energy localized at one single molecule. With a permissible extension to a large ensemble of molecules this would correspond to a high value for the temperature T_1 in (5.27) and result in a high thermal efficiency. The temperature difference between the surface of the sun and the earth is very high so that for thermodynamic processes (photosynthesis) using this energy source the efficiency is nearly unity and the possibility of gaining negentropy is accordingly high. Actually, nature is wasteful here and utilizes only about 31% of the energy that is available. In other cases the thermodynamic efficiency is better, all energy sources present in nature serving to make life processes possible. This is true only in so far as the biological or ecological equilibrium in nature is not disturbed. Thus it is conceivable that microbes might use up the energy stored in fossils, e.g., by the oxidation of the carbon resources such as mineral oil. However, this would result in the using up of the oxygen of the air and hence in the termination of animal life. Of course, the energy carriers produced in plants (cellulose, carbohydrates, fats, proteins, etc.) are repeatedly used by heterotrophic living beings to sustain their own lives but mostly the degradation extends only to a point where the residues can re-enter the biological cycle [47]. Thus, nitrogenous substances in organisms are degraded only down to the nitrates or ammonium compounds.

The supply of free energy to support a nonequilibrium involves a control mechanism (feedback). This is associated with self-instruction. A decisive prerequisite of life is the existence of dissipative structures, i.e., the maintenance of thermodynamic nonequilibria far from equilibrium. This is a dynamic process. Of equal importance is the possibility of transitions from nonequilibria frozen-in in macromolecules to dissipative systems. This is the transformation of a static nonequilibrium into a dynamic, continuous process.

6

Information and information storage

6.1 Fundamentals of information theory

The reception, storage and processing of information from the real world surrounding us is an essential aspect of life. One part of the stored information constitutes the memory, another part furnishes the genotypic information existing outside our consciousness. Part of the processing of information in our brain is called thinking but in the vegetative nervous system and even in the cells information is continually being processed in a variety of ways without our consciousness being involved.

The functions of the brain which are studied and classified by psychologists as thinking, feeling, volition and acting are causally related to material chemical processes taking place in macromolecular cross-linked systems [29]. Thinking consists in forming associations and combinations of information available in words and pictures, giving orders to the locomotor system, controlling courses of actions, etc. The variety of thinking processes is so great that only a few examples can be mentioned here.

It is useful to know some of the fundamentals of information theory [30, 42, 72, 82, 86, 96, 114, 118] if we want to understand the processing and storage of biomolecular information.

We have to make a clear distinction between the carrier and the content of information. Völz [114] speaks of the information carrier and the information carried. In the language of computer technology the term 'hardware' would be employed for the material carrier while 'information content' would be replaced by 'software'. The discussion in this book is almost always about the information carrier. In the few cases where the information content is referred to, this will be expressly stated. In general, information storage and processing are functions of time.

Information means the acquisition of knowledge concerning an unknown problem. In engineering terms this implies the gain in knowledge of a receiver, where the disturbances due to transmission are subtracted from the total content. This question is the basic subject of communication engineering. The message itself conveys knowledge concerning events in space and time. It is transmitted by means of signals which are strictly correlated with other information stored on the receiving side. Thus, in a television set the transferred signals 'black' and 'white' are correlated with fixed spatial and temporal structures by pulses for lines and single pictures so that the 'moving' picture can develop. Between the signals (e.g., 'on' and 'off') and the information there is a coding instruction.

There are four interrelated entities of the coding instruction to be considered:

(1) Information, i.e., information content as pure knowledge, as a command, an instruction, a message, or belonging to the sphere of experience or fantasy;

(2) the conveyer of the message, i.e., the electromagnetic wave, the sound wave, the electric pulse, the electric potential or current, matter flowing in or out, or the stimulation of a nerve;

(3) the path of the message during transmission, i.e., the wire, the air and the nerves, and the location of the message during storage, i.e., the magnetic layer, the photographic emulsion, the television picture, the neurons and biomolecules;

(4) the operating centre for the processing, decoding and evaluation of the information, i.e., the control room, the arithmetic unit and the cortex.

Between (1) on the one side and (2), (3) and (4) on the other there must be an arrangement which we call a code [12]. Entities (2), (3) and (4) depend on one another. In biology, deoxyribonucleic acid (DNA) is the carrier of the messages. Each message is coupled with particular proteins.

As a measure for the value of the information I we define a function of the ratio of the number of possible answers for solving a problem to the number of correct answers after solving the problem:

$$I = f\left(\frac{\text{number of possible answers}}{\text{number of correct answers}}\right). \qquad (6.1)$$

Since the probability P that an event will happen is given by

$$P = \frac{\text{number of correct answers}}{\text{number of possible answers}}$$

we have $I = f(1/P)$. If several cases (answers), 1, 2, . . . , i, having a probability P_i are considered, the value of the information about all cases taken together is equal to the sum of the individual messages:

$$I = \sum I_i.$$

The probability of acquiring this knowledge is an 'as-well-as' probability (the probability for independent, simultaneously-occurring events) and requires multiplicative connection of the individual probabilities:

$$P = P_j P_k.$$

This state of affairs can be formulated mathematically in terms of a logarithmic relation between the information I and the probability P. Then, $I = K \ln 1/P$ or

$$I_1 + I_2 = -K(\ln P_1 + \ln P_2) = K \ln P_1 P_2$$

where K is a constant which is yet to be determined. The value of a message is defined as $K \ln 1/P_i$. The probability of gaining it is P_i. If we receive N messages, each characterized by a probability P_i, we get, for the information content,

$$I = NK \sum_i P_i \ln 1/P_i = -NK \sum_i P_i \ln P_i. \qquad (6.2)$$

Hence the individual message has an expectation value of

$$H_i = KP_i \ln 1/P_i. \qquad (6.3)$$

Normally, the value of a piece of information depends on the subjective judgement of the receiver. Since the information content is connected with particular signs depending on the code chosen and since only the transmitting properties of the apparatus, its speed of transmission and its storage capacity and the signs by themselves can be subjected to mathematical treatment and objective assessment, the mathematical value I is not commonly correlated with the mental content of a message. Only if the information represents a countable selection from many possible cases is mathematical information theory applicable. Therefore, information theory is concerned with the material carriers of information in a storage (e.g., print in a book, semiconductor components in a computer and neurons in the brain) on the one hand and the material paths of messages in transmission (e.g., wires, modulated waves and nerve fibres) on the other as well as with the messages coded into certain signs (coded messages). We have to make a clear distinction between the carrier and that which is carried.

If a message to be received consists of several pieces of information of different probabilities P_i and different values $\ln 1/P_i$ an expectation value

$$H_i = KP_i \ln 1/P_i \qquad (6.3)$$

can be assigned to the partial message. Shannon [96] defines the information value H of a total message to be

$$H = K \sum_i P_i \ln 1/P_i.$$

This formula becomes $I = K \ln 1/P_i$ if the probability is the same for all cases and on the assumption that $\sum_i P_i = 1$ (certainty). The relation $S = k \ln P$ derived in Section 5.3 corresponds to this relation. Thus, the information content I or H can be formally compared to the entropy of a system. Consequently, we can speak of the entropy of information. An increase in information corresponds to a decrease in entropy or an increase in negentropy. At first sight this seems to be a purely formal analogy. It gains much more significance, however, if we assign definite configurations of macromolecules to the storage of information in biological processes. (From the point of view of the theory of reflection in dialectical materialism this means: the macromolecule is the carrier of information and its structure becomes more complex as the complexity of the information increases.) The entropy concept of information is then reflected in the decreasing entropy of the macromolecules that corresponds to this information. In the light of this consideration applying the entropy concept is more than a mere formal procedure.

The formation of a particular molecule from a choice of possible configurations means that one unlikely state is assumed. Thermodynamic probability is identified with the number n of microstates that may be assumed (n being a large number). If there are n possible configurations (microstates of a system) then statistical thermodynamics defines the entropy by $S = k \ln n$. The assignment of a message to a particular unlikely molecular configuration provides evidence that the use of the term 'entropy of a message' reflects the real situation absolutely correctly in the thermodynamical sense and is by no means a formal analogy.

In eqns (6.2) and (6.3) we are still free to choose the logarithmic system of analysis. In communication engineering it is convenient to use alternative statements, such as dead (0) and live (1), or positive (+) and negative (−), or yes and no, or true and false. Therefore, for practical reasons the binary number system is preferred. If there are n different signals the transmission of one correct signal implies the information content

$$I = K \log_2 n.$$

Let us set

$$K = +1.$$

Hence we get

$$I = +\log_2 n \text{ bits/signal, or}$$

$$I = -\log_2 P_i \quad \text{or} \quad H = -\sum_i P_i \log_2 P_i. \tag{6.4}$$

(For $-H$ or $-S$ we use the term 'negentropy'.) The unit of information is called the binary digit (bit) (2^n). When we refer to N bits this means that N signals with a content of one bit each or one signal with a content of N bits are required for complete information. Thus, 16 possibilities require four signals of one bit each ($16 = 2^4$). With a choice from 20 signals, e.g., 20 amino acids, each signal, i.e., each amino acid in the sequence, contains $\log_2 20 = 4.34$ bits of information.

The following scheme represents the transmission of information:

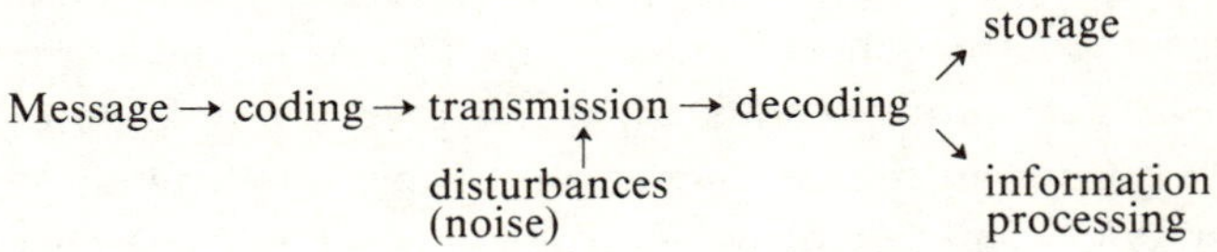

The transmission of information is always an interaction between a transmitter and a receiver of information. Between the two sides there is a channel (line) along which the message to be transmitted may be distorted (generation of noise). The distortions can be eliminated by transmitting the same message repeatedly (redundancy) for it is unlikely that a repetition will give rise to the same distortion.

A distorted message is often useless. Repetition and storage (accumulation) can make it understandable. There is also the possibility of return questions. Our speech contains many redundant elements; most errors are recognized by the receiver and yet the signal is properly understood. Disturbances in a line cover a wide range of frequencies and amplitudes. They can be reduced by confining the message to a small range of frequencies and by repeatedly transmitting and accumulating it. (Moreover, disturbances may be positive and negative and are largely averaged out in this procedure.)

Information may be referred either to the transmitter or to the receiver of a message. In the former case the message is of value to the receiver as long as it has not yet reached him. At the moment of obtaining the information it becomes worthless to him. With regard to organisms this means that the stored information asymptotically approaches a final value during evolution. This final value may correspond to a state of complete saturation, which is equivalent to a thermodynamic equilibrium. The content of information of the receiver grows continuously toward this state and becomes worthless at the same time. On this assumption the information of the receiver may be correlated with his entropy (but not his negentropy). Then even for open systems of living beings the entropy would increase if this were true. This interpretation of the information concept, which is due to von Weizsäcker [117], leads to a contradiction if it is applied to biological storage. For the increasing information content can be stored only in increasingly complex components of the organism. These, however, must deviate more and more from thermodynamic equilibrium. Organisms increase in complexity. As there is a relationship between information and its carrier, evolution leads to increasingly complex and more and more unlikely molecules, i.e., to an increase in negentropy. Therefore, if we emphasize the material carrier of information and consider the analogy between information content and the information carrier, we come to assign the information to the receiver of the message (the second case). In this sense the evolution of organisms towards a higher level means a greater amount of more valuable information for its carriers, a higher negentropy or a lower entropy of living beings. Since living beings are open thermodynamic systems this statement does not contradict the second law of thermodynamics.

6.2 Coding

The signals reaching the receiver are stored, transformed and combined with information already present. In general, the transmission of yes–no decisions requires a convention for the one-to-one correspondence of the signals to the existing information. This convention must necessarily be unique with regard to the direction of the signals (transmitter → line → receiver). The case in question may be the transformation of binary numbers into decimal numbers, of black-and-white information into television pictures, or of modulated radiofrequency into acoustic signals, etc. The prescription for the transformation of one type of signal into another is called a code. Let us consider some examples:

(1) *The binary code*

$0 = 0 \times 2^0$	0
$1 = 1 \times 2^0$	1
$2 = (1 \times 2^1) + (0 \times 2^0)$	10
$3 = (1 \times 2^1) + (1 \times 2^0)$	11
$4 = (1 \times 2^2) + (0 \times 2^1) + (0 \times 2^0)$	100
$5 = (1 \times 2^2) + (0 \times 2^1) + (1 \times 2^0)$	101
$6 = (1 \times 2^2) + (1 \times 2^1) + (0 \times 2^0)$	110
$7 = (1 \times 2^2) + (1 \times 2^1) + (1 \times 2^0)$	111
$8 = (1 \times 2^3) + (0 \times 2^2) + (0 \times 2^1) + (0 \times 2^0)$	1000 etc.

(2) *The ternary code.* This code is similar to the binary code but has so far not found much technological application. It can be realized using, as examples, dc voltage signals +, − and 0 or currents →, ← and 0. This

could correspond to the information yes, no, and unknown, respectively.

$0=(0\times 3^0)$	0
$1=(1\times 3^0)$	1
$2=(2\times 3^0)$	2
$3=(1\times 3^1)+(0\times 3^0)$	10
$4=(1\times 3^1)+(1\times 3^0)$	11
$5=(1\times 3^1)+(2\times 3^0)$	12
$6=(2\times 3^1)+(0\times 3^0)$	20
$7=(2\times 3^1)+(1\times 3^0)$	21
$8=(2\times 3^1)+(2\times 3^0)$	22
$9=(1\times 3^2)+(0\times 3^1)+(0\times 3^0)$	100 etc.

Coding and decoding can be done by mechanical, electrical or chemical processes which can be repeated as frequently as necessary. These processes involve the transformation of continuous variables into discrete steps, i.e., the formation of intervals and approximation by iteration. This implies that there must be a threshold, i.e., a minimum value for recognizing and transmitting information. Thus, analogue quantities are changed into sequences of digits or on–off signals which, in turn, trigger certain mechanical or electrical switching processes. The latter may be transmitted directly through lines or may serve to modulate high-frequency oscillations. Fig. 6.1 gives some examples for this. Appropriate

Fig. 6.1. Digital and analogue transmission of signals: (*a*) Signal; (*b*) signal transformation by step-by-step switching; (*c*) signal transformation into spikes of different densities; (*d*) amplitude modulation; (*e*) frequency modulation; (*f*) received signal after decoding or demodulation.

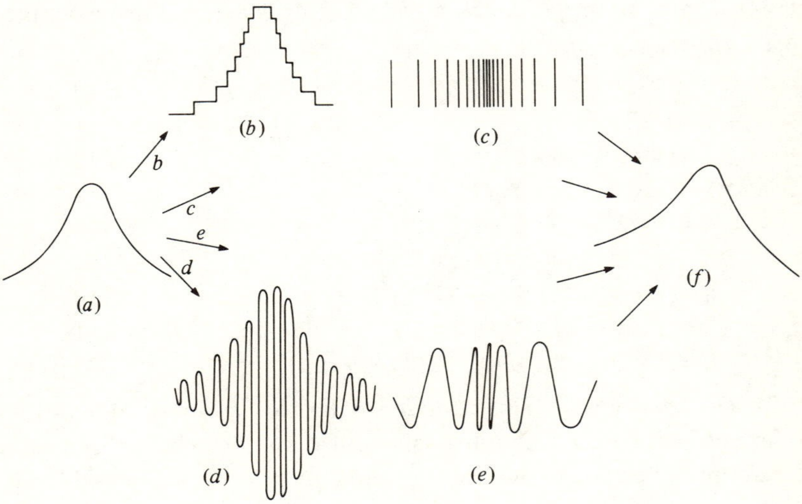

circuits are used for the decoding, e.g., rectifiers for transforming the incoming pulse sequence or modulated high-frequency wave into the original signal.

In certain cases coding is combined with ciphering. This makes a generally accessible message understandable only to selected receivers. For example, in biology this plays a part in the transmission of specific stimuli which activate the sexual behaviour of animals, above all insects, during the process of finding a partner. The stimuli are scents containing ingredients which induce a typical behaviour only in the same species.

In other cases a message becomes intelligible only when combined with some additional information. Thus, in a television, special pulses containing the command to change the line and picture trigger switching processes in the receiver. These switching processes arrange the incoming black-and-white signals in the right position at the right moment to make up the picture.

6.3 Storage [113]

For storage of information in technology, arrangements with at least two stable states are used, e.g., ferromagnetic bodies with north–south and south–north magnetism. Interrogation of a storage should not result in erasure of the information. Otherwise, the information would have to be regenerated after interrogation. Relevant parameters are the storage capacity, access time, storage size, reproducibility and erasability. An important feature is the possibility of transmitting the storage content to a number of receivers simultaneously or successively. By the storage capacity we understand the storage content in bits. The specific capacity is the storage capacity needed per storage element. The access time is the time required to obtain some desired information. Erasability implies that the same storage can be used successively for different pieces of information while the reproducibility of a storage refers to the possibility of repeatedly storing the same message in an identical manner. A list of storage systems, both technological and biological, is given in Table 6.1.

The storage element, which will be influenced by a certain message, such as a yes–no decision, is located at a particular place in the storage. Because it is bound to this location a further information content may be assigned to it. Thus, a particular storage element could carry the information to turn a machine on and off whereas a neighbouring element might increase or stop the supply of material.

The information content is increased if time and place data are additionally transmitted. If these additional messages are missing the information becomes useless or is misunderstood.

Table 6.1. *A survey of storage systems* [113]
(*a*) *Mode of operation*

Type of storage	Recording system	Reproduction (reading)	Erasure	Remarks
Technological storages				
Book	Printing of matter (stencil duplication)	Reading	Impossible	—
Disk	Pressing of matrices	Mechanical/electrical	Impossible	—
Punched card, punched tape	Controlled punching of holes	Optical/pneumatic/mechanical	Impossible	—
Magnetic tape	Magnetic field	Electrical	Alternating field	—
Semiconductor storages				
ROM	Electrical	Electrical	Impossible	ROM = read-only memory
PROM	Complicated electrical systems	Electrical	e.g., UV	Programmable ROM
RAM	Electrical	Electrical	Electrical	RAM = random access memory (semiconductor memory on a bipolar basis) or machine-oriented system (MOS)
Biological storages				
Genotypic storage	Template polymerization	Biochemical	Impossible	DNA
Neural storage	Learning process	Recollection	Forgetting	Networks and proteins

(*b*) *Approximate values for orders of magnitude and switching times of storages*

Type of storage	Atoms/bit	Switching time(s)
Physical limiting value	10^{3}	10^{-13}
Chemical limiting value (DNA)	20	10^{-3}
Proteins	10^{3}	10^{-1}
Video disk	10^{10}	
Photograph	10^{12}	10
Brain (neuron)	10^{13}	10^{-3}
Magnetic tape (video)	10^{13}	10^{-8}
Semiconductor	10^{15}	10^{-10}
Ferrite core	10^{18}	10^{-8}

(*c*) *Approximate values for storage capacities and access times*

Type of storage	Storage capacity in bits	Access time (s)	Size (m^3)
Semiconductor	10^{8}	10^{-8}	10^{-7}
Ferrite core	10^{7}	10^{-6}	10^{-3}
Memory	10^{7}	1	10^{-4}
Human germ cell	10^{10}–10^{12}	2×10^{7}	10^{-15}
Lexicon	10^{9}	10^{2}	10^{-3}
Magnetic tape memory	10^{9}	10^{2}	10^{-4}

7

Information and storage in biological systems

(*General references for this chapter are* [12, 32, 60, 93, 108].)

7.1 Basic concepts of genotypic storage and memory formation [91, 92, 109]

In biology the storing and processing of messages and signals is done by macromolecules; these are the material carriers of the information. We have to distinguish between genotypic storage, i.e., the fixation of the whole inheritance of an individual, and the storage of information acquired in the course of life, in short, the formation of the memory.

The genotypic inheritance concerns the fixation and transmission of characteristics that are typical of, and necessary for, the developing individual, the species, the class and life in general. The genotypic inheritance contains information that is specific for the individual and, in addition, such as is vital for all living beings, animals or plants. It has gradually developed over the last 2000 million years during the course of the evolution of living beings. All genotypic characteristics are stored in each cell of a living being including the germ cell. The extent of the genotypic storage capacity is underestimated and a tremendous amount of vital information is hardly noticed. Raven [88] estimates the inherited information in man to be about 10^{12} bits for each individual. Braitenberg [12] assumes 10^{10} bits. The anatomical structure presented in a textbook of anatomy alone requires $>10^8$ bits of information. In order to give the reader an idea of the power of the set of characteristics to be inherited I give a list which is far from being complete. Estimated values are not given.

(1) The transmission of features that are characteristic of a given individual and may vary within the population of a species. The information for all the characteristics concerned is stored in chromosome segments at the molecular level. The characteristics may be combined by 'crossing over' during the second meiotic

division leading to diploid chromosomes and to randomly-mixed characteristics of the parents. There are an immense number of these features: stature, character, intelligence, susceptibility to certain diseases, talents, inclinations, colour of the hair, colour of the eyes, etc. I leave it to the reader to make an estimate of the wealth of qualities and bodily features which characterize his individuality and which have all come to him by inheritance.

(2) All the characteristics of, and the information for, a huge number of chemical reactions underlying all life processes. These are also stored in the chromosomes. It is known that the life processes constantly going on in the cells are controlled by well over 10 000 different enzymes. The information for the formation of these enzymes according to the genetic code must be fixed in DNA. The same is true for the information about all the very complicated chemical reactions taking place in the cells. Finally let us mention information concerning the formation of low-molecular weight and polymeric substances such as the transport proteins, the storage proteins and the fibrillar proteins.

(3) The information for developing the primordia of all organs and for the efficient working of these organs. Again the genotypic storage must contain this information. Here the sense organs with their characteristic receptors, the organs for all metabolic processes, the sex organs and the primordia for germ cells for the next generation, the supporting apparatus, namely the skeleton and the skin, the muscles, and finally the brain and the neurons are equally important and must be determined by information at the molecular level.

The information mentioned in (2) and (3) above can be considered identical for all individuals of a given species.

The storage system must be unique, indestructible and transferable from cell to cell and from generation to generation. Occasionally-occurring deviations may disappear by frequent repetition of correct information. Also, single errors owing to disturbances are eliminated by redundancy. In unfavourable cases they lead to pathological degeneration. The characteristic genetic information stored in DNA molecules is contained in the male and female germ cells and is transferred through the chromosomes to the next generation in sexual reproduction (transport of genetic information). During embryogeny the biological activity of the genotypic storage contained in all cells is differentiated. The cells develop into specific organ cells and as such they do not require all the information stored in them any longer. The genotypic information fixed in certain localized parts of the chromosomes is inactivated. The cells behave as if it were not present.

Those chromosome parts which retain the ability to transfer their information content to other cells during and after embryogeny are termed heterochromatic. The germ cells for the next generation originate from them. The majority of cells, however, are incorporated in certain organs where they have to fulfil definite functions. Their chromosomes are referred to as being euchromatic. They can only generate cells which are typical of one particular organ. The capability for regeneration fades with growing complexity and increasing age of an individual. The terms 'heterochromatic' and 'euchromatic' refer to different physiological situations but not to new structures with regard to DNA. It is due to different degrees of euchromatization of the same genome that, e.g., a liver cell differs markedly from a nerve or bone cell in its structure and function. Genetically-controlled factors initiating or inhibiting the gene activity may also play a part in this. All organ cells of an individual, be they from the skin, the salivary glands, the brain, the kidneys, or the retina, carry in their dominating DNA structure an identical sequence that is characteristic of that individual. This means that they possess the full information content of the zygote resulting from the fusion of the two parental germ cells.

Since each cell contains the full amount of information it is possible to grow single individuals from single cells by division provided that heterochromatic chromosomes and their capability of regeneration are present. Driesch [21] was able to accomplish this by dividing echinus eggs.

To sum up, all genotypic information forming a living being is fixed in DNA molecules which are present in all cells and have originated from its germ cells. Part of this information supply involves the possible population width of a species while another part concerns all vital functions and all primordia of organs. The information is connected with material storages in the form of macromolecules.

7.2 Acquired information and memory [19, 51, 82, 109, 120, 122]

Of considerably smaller dimensions than the range of genotypic information is that of the experience gathered during a lifetime and constituting the memory.

In assessing storage capacity the concept of a bit is understood in a generalized sense, that is to say, the information value of one bit is assigned to one notion. If, in keeping with the definition given in Section 6.1, we were to attribute an information value of $\log_2 26 \times 8 = 4.701 \times 8 \approx$

Table 7.1. *The acquisition of information by the brain*

	Immediate memory	Short-term memory	Long-term memory
Storage capacity in bits	150	1500	10^6 to 10^7
Influx rate in bits/s	15	0.5	0.05
Influx time	Seconds	20 min	Days
Information durability	Minutes	Days	Life-long
Storage takes place by means of	Networks	Membranes and networks	Protein and DNA formation

37 bits to an eight-letter word we would have higher numerical values for the information content than in the above generalized sense. (As there are 26 letters in our alphabet we have to take $\log_2 26 = 4.701$.) We distinguish between the short-term and the long-term memory; impressions which will be forgotten after a few minutes may be stored in the 'immediate' memory [122]. The access time, which is the time it takes us to become aware of the information, and the processing time are fractions of a second. Events stored in the short-term memory are very quickly forgotten. It can be assumed that all impressions are at first stored in the short-term memory at a growth rate of up to 1 bit/s. Concussions, fevers and cerebral disorders result in a degeneration of the short-term memory (amnesia). Fixation in the long-term memory takes years and the information cannot be erased as a result of mechanical disturbances such as concussions. This is because in the latter case the information has been stored by a chemical process at the molecular level.

If we assume an information value of 1 bit for each notion incorporated into the long-term memory and if we further assume that 1800 new notions are stored in 10 h of learning/day this gives 40×10^6 notions in 60 years. This optimum learning process, however, will never be accomplished so that a storage of 10^6 bits is more realistic [113]. Of course, these 10^6 basic concepts of experience can be considerably extended by forming associations and combinations. The acquisition of information by the human brain, which consists of about 1.5×10^{11} neurons, is summarized in Table 7.1.

The storage of information is connected with a drastic reduction of data. For example, of the 10^6 bits of information produced on a television screen in 1 s ten to twenty bits reach our consciousness. The processing of information begins with unconscious steps followed by the formation of networks in connection with consciousness. Braitenberg [12] gives a good description of these networks in the brain. External stimuli give rise either to motor activity without consciousness (reflex reactions) or to motor activity governed by consciousness and will.

The human nervous system consists of many millions of individual nerve fibres [12, 24, 37, 57]. In the work of Hodgkin [43] and Huxley [50] the conduction of the nervous impulse is assumed to be the wave-like propagation of a depolarization. The normally-existing electronegative state of the nerve fibre (the axon) is interrupted (depolarization) for about 1 ms. The excitation of the ganglion cells after a threshold has been overcome (neuron threshold) gives rise to sequences of impulses (spikes) whose number is a measure of the intensity of the neural excitation. The frequency of the impulses may reach more than 100/s. The electrical depolarization is due to the inward flow of Na^+ ions through the membranes at the surface of the nerve cell accompanied by an outward flow of Cl^- ions (Fig. 7.1*a*). For a short time the action potential rises above zero and thus becomes positive (part A of the curve in Fig. 7.1*b*, steep

Fig. 7.1. (*a*) Saltatory excitation and ionic current normal to the axon direction during excitation. (*b*) Membrane potential (spike) during excitation.

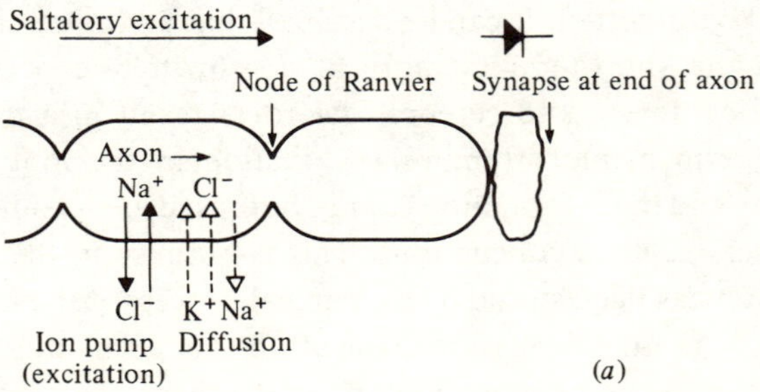

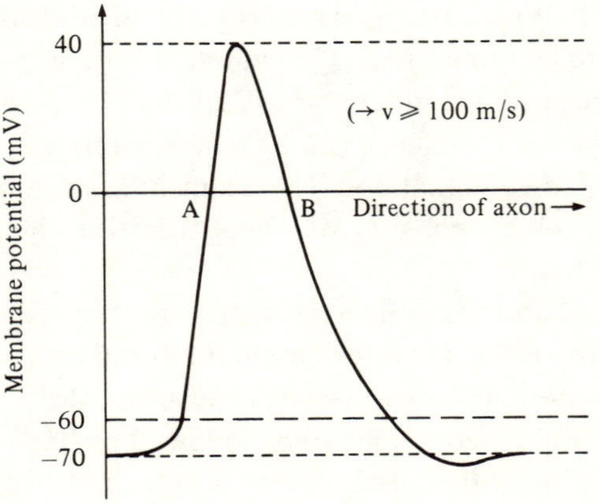

slope). After the neural excitation a slower repolarization takes place (part B of the curve in Fig. 7.1*b*) due to an outward flow of Na^+. Potassium ions are transported in the opposite direction with a slight phase shift (Na^+/K^+ pump). The excitation and the ionic current normal to the nerve fibre move along the axon with a velocity of about 100 m/s. Besides this wave motion the electrochemical excitation can also jump from one node of Ranvier to the next. At the ends of the axons there are synapses, which act as demodulators. The impulses (spikes) can cross the synapses in one direction only (rectification). They cause muscle contractions or generate similar excitations in other nerve cells connected to the synapses. We can regard the excitation of the nerves as coding and the demodulation at the synapses as decoding.

It should be noted here that primarily electrochemical processes are responsible for conduction in nerves [43] and that these are governed by low-molecular weight components (Na^+ and K^+ ions and acetylcholine) as well as macromolecules (in membranes). Thus they are life processes which are beyond the scope of the subject dealt with in this book.

The simultaneous migration of Na^+, K^+ and Cl^- ions through the membrane against the natural concentration gradient takes place as an active transport process with the conversion of ATP to ADP supplying the necessary energy.

In the central nervous system (cortex) numerous connections marked by synapses are integrated into circuits. During excitation, i.e., the formation of action potentials, a great number of possible connections may form which are comparable to the switching of transistors in the forward and reverse directions. The number of possible circuits is enormously large because more than a thousand million switches are involved. The switching processes are the material basis of thinking. They are responsible not only for the acquisition and processing of information but also for controlling the apparatus of locomotion and the sensations, the maintenance of life and the metabolic processes. Circuits either disappear or give rise to a consolidation phase as the electrical potentials of the synapses change and remain constant, respectively. Only in important cases does permanent storage in the long-term memory occur on the basis of chemical processes in the ganglion cells. It is supposed (Zilliken & Abdallah [122]) that the resulting protein molecules facilitate or inhibit the action of definite synapses. Again, an interrelation between DNA and protein molecules may be assumed here.

While the short-term memory is associated with synaptic transmission, the formation of circuits in the long-term memory is connected with the

 Table 7.2. *Concepts of information acquisition*

Sense organs →	Coding	→ Formation of molecular connections	→ Initiation of protein–DNA reactions
Action of external stimuli Excitation of receptors	Formation of electrical potentials moving along nerve fibres at a speed of about 100 m/s In the synapses the nervous impulse is stored electrochemically for a short time	Networks Short-term memory Thinking	Repeated conscious or unconscious storage Long-term memory Formation of DNA–protein complexes Learning Thinking

synthesis of certain protein molecules which are capable of being 'called in' for the whole lifetime of the organism. Recent studies [109] refer also to the role played by lipid molecules in learning processes. Furthermore, the attachment of parts of molecules to existing storage molecules is possible. The parts added are chemically combined with the molecule(s) coding for experience already stored. It should be mentioned that our memory is built up on a logarithmic scale. Thus, events happening in time and space as well as other occurrences seem to be slight and insignificant if they happened long ago or far from us. Presumably, storage is not localized at only one place but occurs in multiple form at different places in the brain. The remembrance of an event or a stored piece of information is called an engram. The currently adopted concepts concerning the acquisition of information are summarized in Table 7.2.

The above results of studies of biological storage in the nervous system, which have been corroborated by experiment, are complemented by fundamental considerations of other authors [67, 68]. Thus, Lorente de Nó [66] speaks of a modification of macromolecules while Fuchs-Kittowski [33] introduces arguments for storage at a higher level. Hassenstein [41] developed the concept of the formation of 'learning matrices' where in each case two neurons crossing each other or connected by synapses are excited. Thus the well-known 'conjunction' arises. This mechanism corresponds to ferrite core storage in technology.

Loose storage in the long-term memory, say by means of hydrogen bridges, seems unlikely since these could be destroyed even by thermal molecular vibrations, very soon resulting in forgetting. Nevertheless, loose storage does have certain advantages. Thus, 'network storage' allows recently-acquired impressions to be discarded rapidly, i.e., irrelevant information can be forgotten immediately after being recorded by the sense organs. This loose storage, which corresponds to weak chemical linkage, can become manifest if it is willed. Therefore, learning does not only consist of consciously acquiring information but very essentially of chemically fixing the latter. If we experience a sequence of unimportant events, e.g., we see a cyclist who ignores the 'give way' sign, crosses a road thoughtlessly and rides on, we will have forgotten about it after a few minutes. If, however, this results in an accident we will be able as witnesses to remember all the events preceding the accident, which, in this latter case, are no longer incidental. The events which, at the time of the accident, were still loosely stored are suddenly recognized as being of importance and fixed in the long-term memory. There must be a macromolecular process running parallel to the psychological process. In technology a similar registration is used in some recording instruments. These operate continuously. The events are recorded on a rotating cylinder and the recordings are erased after each rotation. Only if notable events happen are the recordings transferred from the cylinder on to a paper tape. This transfer also includes the recorded events preceding the one of interest which are still on the cylinder.

There is a further empirical fact which has to be reduced to molecular processes. In the course of life knowledge gained in later phases decays owing to neuron degeneration. The older knowledge, however, is retained. This suggests that storage occurs by successive addition and possibly a successive removal from the end of the storage molecules. We can assume that 'learning' will be interpreted better from a physico-chemical point of view as we get even more insight into the relevant processes.

The chemical changes of neurons and the changes in polarization of the membranes enveloping the neurons are the material components of thinking. 'Remembering' and the reproduction of our knowledge means that our consciousness gets into contact with the storage containing the acquired information. Making this 'connection' sometimes involves some retardation in thinking or requires some impulse from outside. In general, chemically-stored information cannot be forgotten. An old person who suddenly finds himself in a situation where he needs a foreign language he learned in his youth may be surprised to realize that much information concerning that language is still available to him. In the course of life the possibility of transferring information from the short-term to the long-term storage disappears gradually. The chemical fixation of acquired knowledge grows more difficult. This condition is referred to as weakness of memory. Although many events are still perceived, only a part of them is chemically fixed. Time seems to elapse more rapidly. Therefore, linguistic and factual knowledge should be acquired as early as possible; proper reasoning will be learnt later.

Most life processes that establish contacts with the environment serve to acquire information. This particularly applies to the functions of the sense organs which are developed to different degrees of sensitivity, discrimination and reaction rate in the various organisms. Yet information is also obtained by the individual cells, osmotic pressure, the hydrogen ion concentration, electrochemical potentials and active substances such as vitamins being of great importance for this [105].

There are several possibilities for the mechanism of transmission of messages along chain molecules. They are as follows:

(1) A pure electron transport. The conduction of an ionic state along the molecule is predominant. The change of place of the electrons is induced by external fields from the end of the molecule. It corresponds to the mechanism discussed above (Fig. 7.2*a*).

(2) Rearrangements of free radicals along a chain molecule are also possible. This would imply the exchange of an atom or side group at each step (Fig. 7.2*b*).

(3) Finally, σ- and $\sigma-\pi$-bonds can be exchanged along a chain. In each phase of this process side groups would change their places (Fig. 7.2*c*).

In all three cases chemical changes that take place along the molecule can be used to convey information. As was stated above, in biology case (1) is of importance. Changes of place of electrons proceed more rapidly than the inward flow of Na^+ ions through membranes. There remains much to be done in the study of the kinetics and efficiency of these processes.

The stock of experience that we gather constitutes our individuality, our 'ego'. We identify ourselves with the cerebral content of our experience and wish to preserve this storage, which makes our intrinsic nature and is coordinated with our consciousness, for a long time. We wish to live as long as possible and to experience as much as possible. Medical science has succeeded in extending the life-span considerably. At the beginning of this century the average life expectancy in Europe was 40 years while now it is 65–70 years. We have yet to differentiate the notion of age. Three aspects may be distinguished:

(1) The official age, which is valid for all dating purposes in civil life.

(2) The age of the vessels and organs, expressed by the degree of wear in per cent. This is a quantity included in the genotypic

storage. It does not refer to a period of time in the conventional sense.

(3) The age of experience, which will be described below.

Our environment supplies up to 10^5 bits/s of information. Of these, about 15 bits/s are taken in through the sense organs, up to 3 bits/s are stored in the short-term memory and less than 1 bit is transferred to the long-term memory. The most pronounced characteristic of our time is the increase in the rate of information supply by our environment. The rate at which we gain experience has increased very strongly. Our individual development has accelerated (earlier puberty), we have become quicker in learning, we move faster (e.g., in cars and planes), we change

Fig. 7.2. Possible mechanisms of transmission of information along macromolecules: (*a*) electron exchange (change of place of an electron takes place at (*e*)); (*b*) migration of free radicals; (*c*) $\sigma \rightarrow \pi - \sigma$-exchange associated with hydrogen atom exchange in the opposite direction.

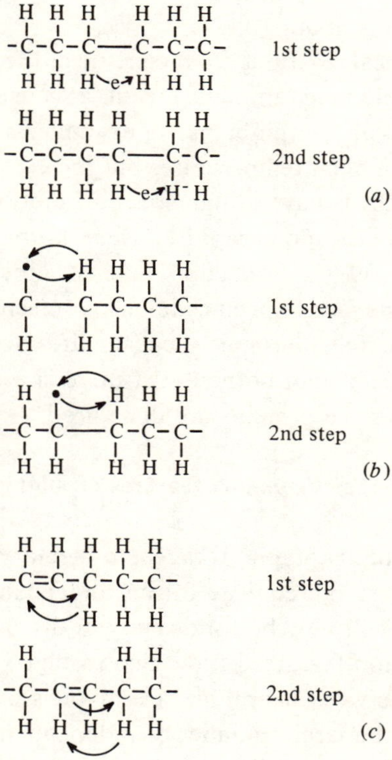

milieux, acquaintances and spheres of action. We are being overfed with experiences from without and within. Never have people experienced so much within one generation. What used to be a rare event is now a matter of course. We are certainly right in assuming a doubling of the rate of gathering experience every 15 years over the last 60 years. This means that in our lifetime we gain an amount of experience which two or three generations ago would have corresponded to a life-span of 300 years. At present the increase in rate approaches a saturation value. Yet it is the most impressive feature of our industrial age. This acceleration of experience prolongs our lives much more than all successes of medical science can do.

I have tried to prove that the conscious gathering and storing of experience is associated with molecular rearrangements. The changes involve overcoming certain potential barriers ΔU between the conformations characterizing the possible states. Thinking and the storage of information in the short-term memory may be supposed to be based on reversible rearrangements in the network of neurons present in the cortex. The rate of rearrangement depends on temperature according to the well-known Arrhenius equation $P = kT/h \exp -\Delta U/kT$. The factor kT/h before the exponential function is about 10^{13}.

The same applies to the chemical reactions characteristic of life. Therefore, the evolution of warm-blooded animals meant a considerable increase in vital activity. Reptiles and amphibians, for example, are active in bright sunlight and at high ambient temperatures but they are almost numbed by low temperatures. Birds have a high speed of motion; they must react to sudden disturbances more rapidly. Their normal body temperature is approximately 39 °C. Protection against illness also requires a higher activity of the body-specific defensive mechanisms. Therefore, we can assume that fever primarily causes life-supporting chemical processes to go on faster than normal. Yet there is a limiting temperature of 42 °C. Above this the proteins are denatured.

7.3 Differences between, and common features of, biological storage systems

The brief descriptions of genotypic storage and memory formation in the preceding sections are prerequisite for an understanding of the subsequent chapters. They will now be followed by a discussion of differences between, and common features of, the two storage systems.

(1) One marked difference between neural and genotypic storage lies in the shorter access time and the larger number of independent pieces of information of the former. But neural storage (in the cortex) is also

characterized by a greater volume or a greater mass as compared with genotypic storage (in the germ cell).

(2) In genotypic storage, all vital information is transmitted from generation to generation without changing the molecular structure. That is to say, the molecular structure as well as the size of the storage is largely retained. In all living beings the same basic information is stored (e.g., for the control of life processes, for biochemical processes, for the conservation of thermodynamic nonequilibria (dissipative systems), for autocatalysis, for self-reproduction). A time of about 10^{17} s, i.e., more than a thousand million years, is recognized for the historical development of genotypic storage. There is only a small increase from generation to generation. Man of 10 000 years ago resembles present-day man very closely in his genotypic appearance. The completion of this storage (evolution) is an extremely slow process.

(3) The memory that we acquire in the course of life, our stock of experience, expands very rapidly. The number of environmental impressions stimulating our sense organs per unit time is the rate of experience. Neural storage ceases to exist at the death of a living being. In man an information content of about 10^6 to 10^7 bits is reached depending on the life-span of the individual. We can, however, transfer our experience to technical storages such as books or tape-recordings. These will remain but we will perish when reaching a degree of wear of 100% or even before this.

(4) It is not possible to fix acquired information in the genotypic storage. Experience gets into the memory (the neural storage) but is not heritable. Memory is a quality attributed to an individual. There is no heritability of data gained by experience.

(5) In numerous cases the same vital information may be fixed either in the genotypic storage or in the neural storage. Thus, we observe that certain abilities that have to be learnt by man are inborn in animals. In the latter case they are called instincts. A new-born chamois can leap in the mountains like its mother. Man must learn to walk in a toilsome process. All the features of a bee's activity are inborn (honey-comb making, care of the progeny, mutual communication, etc.). We are often surprised to find how diversified and appropriate such instinctive actions are. They can hardly be influenced by external stimuli and dangers. They lead to a life proceeding like clockwork, which means a life poor in experience.

In animals, comparatively little experience that is stored in the memory is added to the genotypically-fixed innate impulses and instincts during the course of life. Pavlov [78] showed that animals are also able to store

Table 7.3. *Characteristic data of biological storages (approximate values)*

Type of storage	Time of formation (s)	Storage capacity (no. of bits)	Access time	Erasing time	Remarks
Genotypic storage					
Storage for basic biological information	$\approx 3 \times 10^{16}$	10^8	The genotypic storage of all living beings has developed quite slowly during the history of the earth. The transfer of information to biological functions takes place very rapidly.	—	Parts of the genotypic storage disappeared with the extinction of individual animal and plant species. Individual genotypic storages cease to exist with the death of their carriers; their information contents are transferred to the next generation by sexual reproduction.
Storage not specific to species, e.g., mammal	$\approx 3 \times 10^{15}$	10^9			
Storage specific to species, e.g., man	$\approx 10^{12}$	10^{10}			
Storage specific to individual (embryogeny)	$\approx 2 \times 10^9$	10^{11}			
Neural storage (memory)					
Short-term memory	1	10^4	10^{-1} s	10^5 s	—
Long-term memory	10^{11}	10^7	10^{-1} s	10^{10} s	

acquired information. The well-known achievements in training animals, especially warm-blooded ones, confirm this. Storing all his experience in the memory ultimately makes man free to decide and gives him the freedom of the will.

All those vital actions which are necessary for the survival of the species and for which new-born living beings cannot gather experience are genotypically stored. This applies also to man. The proportion of genotypically-fixed instinctive actions decreases as evolution proceeds and the possibilities of caring for and 'instructing' the progeny improve. During this development the number of actions due to the neural storage rises whereas that due to the genotypic storage diminishes. As life is increasingly governed by experience which is gathered and then stored in the memory, it becomes more and more manifold and diversified. The features of the biological storages are summarized in Table 7.3.

7.4 Storage possibilities in biological macromolecules and their calculation

The question as to how 10^{12} bits of information can be stored at the molecular level is of particular interest from the viewpoint of the physics of macromolecules and will be discussed in some detail.

Each nucleotide found in DNA is characterized by one of the four side groups adenine (A), cytosine (C), guanine (G) or thymine (T). There are four different monomeric nucleotides and $4^2 = 16$ different diads, namely, AA, AC, AG, AT, CA, CC, CG, CT, GA, GC, GG, GT, TA, TC, TG, TT. For $n = 500$ this makes 4^{500} or about 10^{300} chemically different DNA molecules. The universe would not be large enough to hold one specimen of each of these different DNA molecules made up of 500 basic molecules each. Hence there are so many chemically different biomolecules that apart from identical twins and living beings produced parthenogenetically or by gemmation we can assert that there have never existed two individuals generated by normal sexual reproduction whose polymeric DNA molecules have been chemically identical.

How can the largest possible amount of information be stored taking advantage of this huge number of possible, chemically different macromolecules?

We have to distinguish between dependent and independent information. In the case of dependent information it is possible to store or transmit several connected information contents by means of one signal. In other words, dependent processes are those which always occur together and are biologically meaningful only in their totality. Thus, processes such as photosynthesis, glycolysis, the Krebs cycle, etc. are composed of a great number of simultaneous events. Therefore, a small number of independent messages linked by numerous pieces of dependent information are sufficient to transmit such a 'chemical programme' to a subsequent generation, say through the seeds of a plant. In other cases seemingly quite different independent characteristics are interrelated genotypically. Thus, albinos will have white skin, red eyes and other typical features. In many hereditary diseases several organic disorders occur simultaneously; this indicates that they are due to the multiple utilization of certain storage possibilities.

If a storage consists of n^2 elements, e.g., a square matrix of n rows and n columns with an information content of α bits/storage element it can be used to store $n^2\alpha$ bits independently. In DNA α is the information content in bits per purine or pyrimidine base. Here each of the four possible bases A, C, G and T has an information value of 2 bits if the position in the molecule relative to the beginning or to some starting place is also given. A correlated protein molecule with n amino acid residues has a storage capcity of $n \log_2 20 = 4.32\, n$ bits/protein molecule for independent information. Maximum molecular lengths of $n = 10^9$ nucleotides have been observed. This corresponds to maximum storage capacities of 2^9 bits/DNA molecule. For correlated protein molecules with a degree of polymerization of $n/3 \approx 10^9/3$ we would obtain $4.32/3 \times 10^9 = 1.44 \times 10^9$ bits/molecule. This kind of storage supplies independently-ordered information contents.

If Z bits of information have been stored and all elements determined with respect to their sequence and spatial position each new independent information content requires a new element. It is not possible to use the elements present for further messages without disturbing the information that already exists. Thus, e.g., an adenine group at position 743 of a DNA molecule might determine the colour of the skin of the individual carrying this DNA. Interrogating the sequences over such a long molecule will lead to incorrect results even if a mistake is made in one place only, e.g., if one base group is left out. Since in such a case the wrong proteins would be synthesized because of the genetic code this would have unimaginable consequences. The possibility of incurring such errors in interrogating a molecule along the strand is pointed out by Jung in a fundamental paper [54].

However, a large amount of dependent information can be stored in any one DNA molecule. Determining the base sequence, e.g., ACCTGAAGGA also fixes the successive diads, in this example AC, CC, CT, etc. as well as all the triads, namely ACC, CCT, CTG, TGA, etc. Altogether there are $n(n-1)/2$ diads, $n(n-1)(n-2)/6$ triads, etc. As is easily seen this is a total of 2^n polyads. For 500 groups A, C, C,

T, G, etc. this amounts to 10^{150} bits of dependent information stored.
This greatly exceeds the number n of independent bits of information stored. It is probable that the storage arrangements discussed so far are of only minor importance in the biological sphere because reading errors may easily occur.

Storage in characteristic sequences (polyads) of the nucleotides is less prone to errors and more unambiguous than storage in smaller elements. The bases A, C, G and T and the 20 different amino acids may themselves be the carriers of information but it is more probable that only the n^2 diads AA, AC, AG, . . . , the n^3 triads AAA, AAC, . . . , etc. can accomplish this task. The total storage capacity is given by a geometrical series, $n+n^2+n^3+n^4+\cdots+n^m$, which is equal to

$$[(n^m-1)/(n-1)]n. \tag{7.1}$$

The resulting code is based on assignment to the polyads. Each polyad, ADT, GCGAA, etc. (corresponding to a word in a language) would then be the carrier of one biological message. If n is the number of the different bases in DNA, m represents the influence among neighbours. If $n=20$ (*proteins*) *and* $m=8$ (octads), which means that at most 8 neighbouring nucleotides influence and recognize one another, then 2.69×10^{10} different and independent pieces of information are possible.

Modern nmr spectroscopy provides evidence concerning structural details of chain molecules over a few monomeric units. Using very sensitive equipment one can detect at most six neighbouring groups with this technique. We may assume that biological processes display a similar limiting sensitivity.

In order to avoid overlap, the individual polyads carrying discrete pieces of information must be separated from one another by special marks. The above-mentioned storage principles can be combined in such a way that, for example, the chromosomes contain DNA molecules consisting of p polyads whose order is fixed and which influence one another. Each polyad is made up of a maximum of m nucleotides. Then the storage content is calculated to be

$$p(n^m-1)n/(n-1) \text{ bits/storage element.} \tag{7.2}$$

With $p=100$, $n=4$ (DNA), $m=8$, we have 8.7×10^6 independent pieces of information. This number is considerably increased if permutations of the p different polyads are permitted and a new information content is assigned to each permutation. This type of storage is similar to the formation of human language.

The formation of words and sentences is an optimum system as regards avoiding errors and the number of redundant elements. Thus, a sequence of letters, such as that in the word 'together' stores a number of other notions, namely, 'to', 'tog', 'get', 'eth', 'ether', 'the', 'he', 'her'. Several words constitute a sentence corresponding to the *p* polyads of DNA. Rearrangements of words and letters (permutations) result in different information contents. Take, for example, 'bathwater' and 'water-bath', or 'May I come tomorrow?' and 'I may come tomorrow'.

In a spoken language certain words are excluded because they would be too difficult to pronounce. Other sequences of letters are interdependent. Thus, in the English language the letter *q* is always followed by *u*. Similar regularities in the base sequence A, C, G, T in DNA are possible but not yet known. Some observations suggest that base sequences in one strand are mirrored in the neighbouring strand of DNA, e.g.,

...AACGCTTA... (molecule 1)

...ATTCGCAA... (molecule 2).

Points of base sequence inversion could indicate the beginning of a message. In human language individual words are combined to form a sentence in which several pieces of information are related to one another. Sentences express the interactions between subjects and objects. They mediate an immense variety of interrelations, conditionalities, qualities and actions with respect to space, time and motion. Similar connections of the macromolecules carrying information are supposed to exist in biological processes.

Fertilized germ cells are the most complicated and compact storages. The whole of the genotypic information consisting of more than 10^{11} bits is stored within one single cell. Assuming a hundred-fold storage of the genome (safety factor of 100) this requires a total of about 10^8 macromolecules with a total mass of about 10^{-6} g. The whole molecular storage can easily be contained within a cube having an edge of 10 μm. Many of these messages must be independent of one another.

7.5 The entropy of biomolecules [106*a*]

Thermodynamic quantities refer to large ensembles of particles of the same kind, e.g., to 1 mol, 1 kg, or 1 m^3. The question arises as to whether thermodynamic considerations can be applied to a single molecule. This is possible if we use information theory and the entropy concept as defined by it [96]. Let us consider two examples:

(1) Let a DNA molecule contain N_1, N_2, N_3 and N_4 bases of A, C, G and T respectively. If we assign the information to a quantitative distribution of the bases that is independent of the base sequence in the molecule the numbers of molecules determined by the fractions N_i/N represent the groups containing the information in the case assumed

here. A particular base distribution for N given nucleotides has

$$N!/(N_1!N_2!N_3!N_4!)$$

molecules. (Base permutation does not yield differently-structured molecules.) The probability that any of the 4^N possible molecules with a polymerization degree N belongs to the fractions determined by the given N_i is

$$P = \frac{N!}{N_1!N_2!N_3!N_4!4^N}. \tag{7.3}$$

For the natural logarithm of the probability we obtain

$$\ln P = \ln N! - \sum_i \ln N_i! - N \ln 4.$$

Using Stirling's formula we get, after some intermediate steps,

$$\ln P = -N \sum_i N_i/N \cdot \ln (N_i/N) - (\tfrac{3}{2}) \ln N - N \ln 4$$
$$- \tfrac{1}{2} \sum_i \ln (N_i/N) - 3 \ln \sqrt{(2\pi)}. \tag{7.4}$$

If as a special case we set $N_1 = N_2 = N_3 = N_4 = N/4$ we have

$$\ln P \approx -\tfrac{3}{2} \ln N.$$

For the information content H with $P_i = N_i/N$ a comparison with Shannon's formula yields

$$H \approx \left(-N \sum_i P_i \ln P_i - (\tfrac{3}{2}) \ln N - N \ln 4\right) \Big/ \ln 2 \text{ bits/molecule} \tag{7.5}$$

or, in the special case of $N_i/N = \frac{1}{4}$, $H \approx \frac{3}{2} \ln N/\ln 2$ bits/molecule. In the first example the fractions N_i/N of the individual bases determine the molecular arrangements compared with one another. With increasing degree of polymerization the negentropy rises comparatively slowly.

(2) If we consider the information to be determined by the base sequence in a DNA molecule then one of the 4^N possible molecules will carry the correct information. The probability of encountering such a molecule is 4^{-N}. As there are 20 different amino acids, each of them determined by 3 nucleotides according to the genetic code, we get $P = 20^{-N/3}$ for a protein molecule correlated with the DNA molecule under consideration. With $N = 500$ this means that the probability is $4^{-500} \approx 10^{-300}$ for DNA and $20^{-166} \approx 10^{-200}$ for the protein. These are enormously low probabilities. They are associated with a very high negentropy. The probability that this particular molecule is not present is very much higher, namely, $1 - 1/4^N$. Substituting these expressions into

 Shannon's formula gives

$$H = -\sum_i P_i \frac{\ln P_i}{\ln 2}, \tag{7.6}$$

$$H = -\frac{1}{4^N}\frac{\ln(\frac{1}{4})^N}{\ln 2} + \left[1 - \left(\frac{1}{4}\right)^N\right]\frac{\ln(1-(\frac{1}{4})^N)}{\ln 2},$$

$$H \approx \frac{2N}{4^N}, \quad \text{so } H \approx 10^{-300} \quad \text{for } N = 500.$$

A polymeric DNA molecule with a definite base sequence has a very low entropy in the sense of information theory and hence a very high information value.

The replication (reduplication) of a storage molecule by template polymerization does not violate the fundamental laws of thermodynamics. The reduplication of the unlikely base sequence does not require free energy. Systems of very high negentropy, i.e., of high information content, can be reduplicated. The information is multiplied and transmitted in an optimum way. The conservation and reproduction of the thermodynamically unlikely biological storages can be regarded in just the same way as the production of a book, which is never done by putting letters together at random but by setting type according to definite rules and prescriptions. The printing machine produces duplicates of the given manuscript in a very simple manner. In the cells the information-carrying molecular storages are replicated in a similar way. As evolution goes on the genotypic storage grows more and more complicated, its negentropy (information content) slowly increasing from generation to generation.

Macromolecules formed by template polymerization, which conserves information, can remain in their original configuration even without showing appreciable signs of active life. Only macromolecules freeze-in in a state far from thermodynamic equilibrium and show no tendency to change into another state which would falsify the information.

Macromolecules are destroyed, and their information content is annihilated by heating, the absorption of quantum radiation and bombardment with fast particles. In a frozen-in state biomolecules retain their basic information for a long time and may originate a biological era on any planet having an environment conducive to life. The most important information in this regard is that for establishing dissipative states (flow equilibria) and self-reproduction. I have given an extensive description of the conditions necessary for such a process elsewhere [47]. This possibility will be discussed in Chapter 11.

7.6 Errors in the replication and translation of biomolecules, particularly in mitosis and meiosis

Besides purely random errors occurring during template polymerization deviations and sequence rearrangements occur in biomolecules as a result of irradiation, local heating and chemical reactions. Many of these changes do not lead to the complete destruction of the affected molecules but to structural changes or to the degeneration of biomolecules which then undergo further replication. In this context we should mention in particular hereditary diseases and perhaps the origins of carcinomas. However, not every disturbance of a molecule results in a pathological cell degeneration or a hereditary disease. This would require the same type of damage to many neighbouring molecules.

Smaller energies are required to disturb the purine–pyrimidine linkage than are necessary to break up biomolecules themselves. Tertiary and quaternary structures are easily altered by thermal vibration. The denaturation of proteins is observed at temperatures above 42 °C. The ultraviolet (UV) radiation from the sun is sufficient to destroy biomolecules. The process in question is a skin damage which is to be understood not in thermal but in quantum-chemical terms. It occurs even at room temperature. Since the radiation is absorbed by the skin, with the human body protecting itself by the formation of pigment, the skin is damaged (sunburn) only if subject to very intense UV radiation. Ultraviolet, and, even more so, X-ray irradiation cause cells to be killed and errors in replication and translation to occur (leading to skin cancer?). Of course, ionizing radiation (e.g., in radioactive decay) penetrates into the tissue much deeper and causes the decomposition or cross-linking of molecules.

All living beings are constantly exposed to cosmic radiation, which gives rise to molecular rearrangements in the body. The radiation load due to the utilization of nuclear energy and the working hours spent at nuclear power stations is very much smaller than this.

Let us now discuss errors in the formation of biomolecules and their elimination from a stochastic point of view. According to Eigen [25, 26*a*] an error rate of about 1% with respect to the base sequence of the total molecule can be expected. This means that one error occurs at some point of a molecule per one hundred error-free sequences. One such error in the correlation of the purine and pyrimidine bases or in transcription and translation will lead to one wrong message and will recur in subsequent replications. It is possible to reduce the error rate by multiple storage, i.e., by redundancies. A multiply-transmitted message is unlikely to be affected with the same error each time. The crossing over during the second meiotic division mixes the information descended from the parents randomly. Probably this happens in such a way that the male as well as the female share exists as a multiple storage in the exchanged segments of the chromosomes. If we denote the right and the wrong information by I_r and I_w respectively, and the probability of transmitting the latter by $a\%$, we have an error rate of $a\%$, say 1%, in single transmission. For n-fold transmission, using the binomial theorem, we calculate the amount of wrong information occurring to be

$$I_w = (a/100 I_w + (100-a)/100 I_r)^n.$$

Hence, for 100^n possible cases, per n transmissions we have a^n cases with n-fold error copy; $na^{n-1}(100-a)$ cases which are $(n-1)$-fold wrong and $n(n-1)a^{n-2}(100-a)^2/2!$ cases which copy the error $(n-2)$-fold. For $(n-m)$-fold error copy this yields $\binom{n}{m}a^{n-m}(100-a)^m$ cases (m and n are integers). If only that information for which $(n-m) > n/2$, i.e., $m < n/2$, is replicated incorrectly, the error rate α_w is given by

$$\alpha_w = \sum_{m=1}^{m=\frac{1}{2}n-1} \binom{n}{m} \frac{a^{n-m}(100-a^m)}{100^n}.$$

In the case of three-fold storage the error rate falls from 1% to 2.98×10^{-4} and with five-fold storage it diminishes to 0.985×10^{-5} if we set $a = 1\%$.

This one example of a calculation may suffice to show quite impressively to what extent the error rate is decreased by redundancy, i.e., by multiple storage of the same information. On the other hand it is obvious that because the information is recorded from the beginning of a molecule during information transmission, on average an error is already incorporated after $100/a$ sequences. This makes all the subsequent information of the long molecule wrong because it is affected by the error.

8

Living beings and computers or robots

8.1 Fundamentals of computers

Organisms are comparable to automata, i.e., robots, but also to electronic computers in some respects. (Robots and computers cannot always be clearly distinguished from each other conceptually.) The comparison could also be attempted with higher organisms and man. In all cases information from outside is taken in and uniquely processed in the interior. Logical correlations are produced that result in an outward response. The logical systems correlated with definite instructions and inputs are called algorithms. Perhaps the experience gathered on electronic machines and their possibilities help us to interpret thinking processes and to add to the knowledge of comparative psychology.

Before we can make a comparison between the processes going on in automata and life processes we must understand the mode of operation of these computers. Mostly they are program-controlled computing machines which interconnect a certain sequence of numbers on logical principles accomplishing an algorithmic scheme. They may make the output data obtained pass the given program repeatedly without appreciable intervention from outside. They operate either as digital (from Latin *digitus* = finger) or as analogue computers. In the latter case they convert the given number sequence into analogous quantities, say geometric or electrical ones, which they vary in a predetermined way. The automaton receives certain instructions concerning the treatment of the number sequence to be processed. The numbers entered are at first changed into electrical pulses and then passed on to the arithmetic unit of the machine via a control unit. The instructions and partial results are also stored in the machine, that means, they are passed on to the register. All the calculations are done according to a program, step by step, at exactly prescribed intervals of time. The register contains the addresses (individual storage cells) and represents the 'brain' of the machine.

Mostly, there is no difference between the storage for the data and that for the instructions. Every computer carries out fundamental arithmetic operations, stores the results and can, if certain conditions are reached, at a given point either output the result or pass it through the cycle again. Punched cards, punched tapes, magnetic tapes and magnetic core memories are used as storages. The development of microelectronics has given rise to microsystems in the form of magnetic thin-layer storages, microprocessors and digital plotters. The individual storage cells are the addresses; each instruction has an address assigned to it. The addresses are processed following a flow scheme, which is the flow diagram of an algorithm. An algorithm is a unique instruction for solving a number of equivalent problems (e.g., extracting roots of positive numbers).

Consider this example of an algorithm:

(1) Store a given number x in R_1, move it to R_2 and add to it the number y stored there; $x+y=z$.
(2) Divide the number x stored in R_1 by the number y stored in R_2, move the result to R_3.
(3) Subtract the number z from this and, if the result >20 return it to R_1 and there add it to x, otherwise stop.

An important point in each program is the possibility of returning to previous process stages. This is absolutely analogous to the self-controlling feedback and the establishment of flow equilibria by means of feedback mechanisms in organisms. In computer science a mathematical logic has developed, the basic elements of which are negation (not), conjunction (and), disjunction (or), alternative (at least one), implication (if . . . then . . .) and equivalence (if and only if . . .).

A computer always works on the basis of a given program stored in a register. The automaton receives the instructions in the 'language' specific to it or converts them into the latter. In addition to numbers all physical quantities can be put in or automatically received in response to instructions. Optical impressions are received via photoelectric elements, acoustic impressions by means of microphones, thermal impressions by electric thermometers, pressures by manometers, electrical quantities by means of moving-coil instruments and watt meters, etc. Transducers change these quantities into numerical values for digital processing or into analogous electrical quantities. The resulting electrical pulses change the polarization of ferromagnetic storages or, after appropriate amplification, actuate relays which print out the results or punch cards or tapes. The networks which are joined to the arithmetic

units and carry out the logical operations represent the 'machine brain'.
Machines work extremely rapidly. They perform a great number of operations in fractions of a second.

8.1.1 *How does a computer work?*

First the human operator of the machine determines the extent of the problems to be solved. He transcribes these into the language of the machine and designs programs which are optimized. This means that the calculating procedure is uniquely determined with a minimum of necessary steps and also the permissible calculating time is fixed. Every machine can run a variety of programs depending on its storage capacity (single and multi-address machines). The development of the machine requires very extensive designing preceded by a thorough consideration of all possibilities, and the exclusion or blocking of any wrong decisions of the machine.

8.1.2 *What can a computer do and what can it not do?*

A computer can solve problems and execute all the programs that are entered into it. It can store relevant or irrelevant results during the first run, in which many steps proceed in the 'or' and 'and' modes. These results can be used again in subsequent runs. The machine can 'remember'. It learns and takes decisions according to the rules it has accepted. It remembers wrong decisions and avoids these in repeating the cycle. It uses pathways for its calculations and automatically chooses the fastest to save time. In other cases a robot recognizes its 'master' when he sets a prescribed number which he alone knows, e.g., it gives the alarm (protection against thieves) if a stranger approaches it or illicitly crosses a barrier. It never uses its own discretion. It understands only instructions in the machine language, i.e., the algorithms of its programs. These avoid uncertainties and ambiguities which are inherent in natural languages. The machine can repair itself if the designer has made provision for this, e.g., by installing standby units which are automatically switched on if the main arithmetic unit breaks down. Deviations from the program are possible unless the designers have provided for an appropriate restriction or the self-destruction of the machine. If the latter does not happen the robot may become the master and the designer the slave (see Goethe's poem *Der Zauberlehrling* (*The Sorcerer's Apprentice*)). Finally, the computer can design calculating programs itself if the electrical networks are refined to such an extent that only a few commands are sufficient to set up and run a program. The robot could reproduce itself if it were equipped with the appropriate tools and material (automatic production of automata!). This, however, has not yet happened.

8.1.3 *What are the limitations of a robot?*

(1) A robot can only perceive objects, measure their characteristic data and process these comprehensively, for which task it has been equipped by the designer with the appropriate sensors. A robot with photoelectric cells and counter tubes designed to detect light and X-ray signals will not respond to acoustic signals.

(2) A robot will not recognize itself, only objects of the outer world. Photoelectric elements may be able to record the robot's own equipment but they cannot apprehend the computer design as a whole.

(3) A robot will not recognize creative man as an intellectual being, only certain material peculiarities for which it has sensors.

(4) A robot will never 'think over' the questions of whether it has been created and who it was that created it because these questions are not included in its program.

(5) An automaton does not originate by chance, say, e.g., as a result of putting together and thoroughly mixing in a box glass components, ferrite powder, pieces of wire, etc. Therefore, we cannot, in the case of the much more complex living beings, speak of an accidental origin here either.

(6) Recognition is definitely limited in an automaton and we are ready to concede this to it. The automaton itself does not 'recognize' this limitation. It is 'convinced' that it knows everything, stores everything very rapidly and processes everything correctly from a logical viewpoint. As far as it is concerned, there is only that which it can recognize; everything else does not exist for it and it does not allow any questions relating to objects beyond its limits.

(7) An automaton only knows yes-no decisions. Intermediate stages and emotional statements are beyond its possibilities.

The automaton is suitable for a study of the limits of human understanding. Recognition of the fact that there is also such a limitation for us requires the transfer of this knowledge to organic, in particular to human, life. The similarity between the automaton and organic life is more than a mere external correspondence. We should always remember that ultimately man is nothing but a super-automaton whose design and variety cannot be comprehended [31, 101]. Let us further note that no automaton originates on its own, by chance. Any automaton understands and processes only that which it has been programmed to do by its designer from the very beginning.

Table 8.1. *Comparison between man and a computer or robot*

	Automaton	Man
Contacts with the outer world	Yes, sensors for all physical quantities possible	Yes, sense organs
Information storage, memory	Yes, core memories up to 10^9 bits, magnetic tapes, thin-film memories	Yes, (1) genotypic storage up to 10^{12} bits, (2) storage of acquired knowledge up to 10^7 bits, long access times
Formation of feedback	Yes, repeated input and feedback for certain calculation of results	Yes, feedback, flow equilibria, auto-catalysis, closed control loops
Logical connections	Yes, rapid and unique, decisions are fixed by prescribed algorithm, no emotions, only yes–no decisions	Yes, many-sided freedom of the will and freedom to decide, yes–no decisions and intermediate solutions, emotions and sensibility, high mental flexibility
Mental abilities with respect to the freedom to take decisions	No	Yes
Learning processes	Yes, fixation of pathways in repeatedly passing the cycle, storage of the path found to be optimum	Yes, learning and storage, formation of correlations
External energy source	Yes, external energy (electrical energy)	Yes, food (chemical energy); entropy production, conservation or decrease of the entropy of his own open system
Acquisition of further information	Yes, input of additional information in free addresses	Yes, learning processes, evolution
Taking decisions in unforeseen situations and independent extension of program	No	Yes (evolution)

Table 8.1. (*cont.*)

	Automaton	Man
Evolution	More and more advanced automata are constructed by man. Self-improvement has not been possible so far	Yes, without recognizable intervention from outside, selection on the assumption of selection pressure
Self-healing	Yes, possibly electrical branches or standby devices which are automatically switched on in the case of a breakdown	Yes, regeneration
Reproduction	No, only new making; self-constructing automata have not yet been devised	Yes, natural sexual reproduction
Origin	Never by chance. Only creative man designs and constructs automata. Any automaton is based on a man-made functional diagram. It is impossible to make a computer by putting components together at random	The random origination of life in a dead environment would be perpetual motion of the second kind. Probably it is frozen-in flow equilibria that initiate dissipative (dynamic) states. Orientated random processes are also conceivable (see Chapter 11)

8.2 Comparison between man and a computer or robot

In trying to understand a computer we should consider two aspects:

(1) The material carriers, networks, storages and their construction (hardware);

(2) the program, the information and the logical interplay (software).

The comparison made in Table 8.1 shows the similarity of an automaton to a living being. Yet the table also makes clear the immense advantage of all living beings. Biological storages are extremely small but they have an enormously large capacity. The human germ cell is estimated to have an information content of about 10^{12} bits. There are innumerously many connections between neurons. There are many intermediate stages between 'yes' and 'no', there are ethical and aesthetic sentiments, feeling and volition. It should not be overlooked, however, that man-made automata react very much faster than do biological systems. The fundamental differences between living beings and automata (created by man and not originated by chance) outlined here are regarded as insignificant by Steinbuch [101]. On the one hand, life functions are understood better and better and their similarity to automata is elaborated more and more clearly as our knowledge increases; on the other hand, automata are made more and more similar to the natural prototype.

However, it remains absolutely incomprehensible how decisive life phenomena such as the acquisition of consciousness, the freedom of the will and the growth of human knowledge could be transferred to automata. Human life would be quite poor if morality, pleasure and all the enjoyments of art were to be understood as algorithms and programmed experience. Programmed (fatalistic) life would assume a programmer. The concept of 'the freedom of the will' would then be meaningless.

9

Regulation and control in biological processes

9.1 Fundamentals of regulation (feedback control) [46, 76]

Regulation serves to keep certain measurable quantities constant or, following a program, to vary them. Regulation and control are subdivisions of cybernetics. I shall restrict discussion to control actions at the molecular level, in the cells and the living organism as a whole.

Feedback control makes use of closed loops consisting of the controlled process, the measuring element, the controller and the control element. The output quantity is constantly being measured, registered and monitored and compared with the desired value. In open-loop control only certain commands are introduced into the process without measurement or checking of their effectiveness. The principle of closed-loop control lies in recognizing a deviation x of the actual from the desired value, comparing it with the desired value w, processing and transforming the deviation x and intervening by means of a correcting value y in order to remove a disturbance z. Sometimes the trend of the deviation, i.e., the rate of change of the control quantity, $\mathrm{d}x/\mathrm{d}t$, is monitored. Often deviations below a threshold value Δx are permitted or an intervention proportional to the deviation x is made. Disturbances can be removed even more efficiently by integrating over the time in which the disturbance influences the control quantity $x(t)$. The correcting value y grows proportionally to the integral $\int x \, \mathrm{d}t$. To compensate a disturbance z the controller must change the correcting value $y(t)$ in the reverse direction, i.e., opposite to the disturbance. The resulting control loop is shown in Fig. 9.1. x, y, z and w are functions of time. Mostly $z(t)$ or $w(t)$ is given while $x(t)$ has to be found.

In control engineering linear functions are used. Small deviations from the desired value occur. Using Taylor's series expansion all physical conditions can be linearized. In passing through the controller the control quantity $x(t)$ is transformed into the correcting value y_o ($y_o = f_1(x)$). At the controller output phase reversal takes place ($y_i = -y_o$). The correcting value y_i and the disturbance z influence the controlled process. We have $x = f_2(y_i, z)$ as the equation for the controlled process. The interconnection is made in such a way that despite the disturbance z the control deviation x is as small as possible. An excess of energy and matter will ensure that any occurring deviation is eliminated.

From the equations

$$z + y_i = s_0 x_o + s_1 \frac{dx_o}{dt} + s_2 \frac{d^2 x_o}{dt^2} \quad \text{(controlled system)}$$

and

$$y_o + T_1 \frac{dy_o}{dt} = r_{-1} \int_0^t (x_i - w)\, dt + r_0 (x_i - w) + r_1 \frac{d(x_i - w)}{dt} \quad \text{(controller)}$$

with

$$y_i = -y_o \quad \text{and} \quad x_i = x_o = x$$

we obtain, after some calculation,

$$z + T_1 \frac{dz}{dt} + r_0 w + r_1 \frac{dw}{dt} + r_{-1} \int_0^t w\, dt = r_{-1} \int_0^t x\, dt + (r_0 + s_0) x + (r_1 + s_1 + s_0 T_1) \frac{dx}{dt} + s_1 T_1 \frac{d^2 x}{dt^2} \qquad (9.1)$$

Fig. 9.1. Block diagram of a control loop consisting of the controlled process (e.g., organ or chemical reaction to be controlled), the controller (e.g., enzymes), the measuring element (e.g., enzymes, chemical affinities) and the control element (e.g., energy supply).

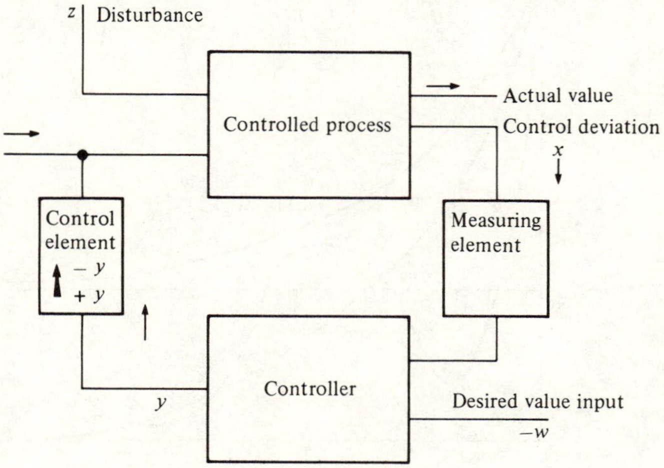

 where the following constant coefficients are derived from the physical and structural features of the set-up: r_0 provides the correcting value y_o which is proportional to the control quantity x_i (p-action controller), r_1 is of importance if the controller has a derivative-action component (d-action controller) ($y_0 = r_1\, dx_i/dt$), r_{-1} occurs in integral-action controllers (i-action controllers) ($y_o = \int x_i\, dt$), T_1 characterizes delayed action of the controller, s_0 provides the correcting value y_i proportional to the control quantity x_0 ($y_i = s_0 x_0$) and s_1 refers to first-order lags within the controlled process. Controllers mostly have self-regulation, which is expressed by $s_0 \neq 0$. In this equation a reference or desired value $w(t)$ passing through the controller may also be of importance.

Eqn (9.1) describes a reaction $x(t)$ which, with a given disturbance $z(t)$, gradually becomes $x = 0$ or $x = x_\infty \neq 0$. For large x, y, w and z, there occur deviations from linearity which cause the final value to be constant or to fluctuate periodically. Usually a disturbance z is assumed to occur

Fig. 9.2. Transient response for a closed loop with a given constant disturbance z at time $t = 0$. Curves: (*a*) control without integral term (remaining control deviation) (eqn 9.5); (*b*) control with integral term (complete elimination of the disturbance) (eqn 9.6 or 9.7); (*c*) unstable behaviour (oscillator with oscillations building up and then remaining constant).

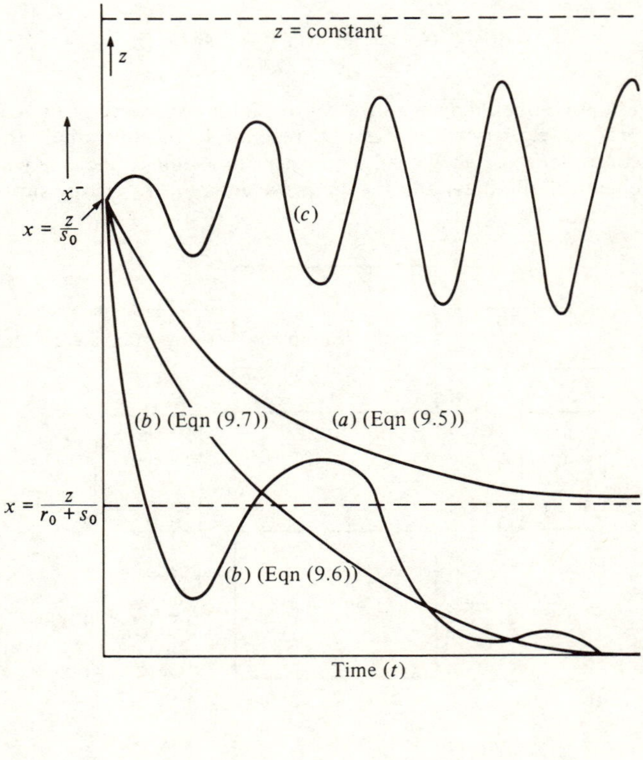

as a jump at the time $t=0$ and then remain constant (jump function).
The resulting time-dependent control deviation $x(t)$ or the deviation of the correcting value $y(t)$ is called the transient response (Fig. 9.2). Since in practice the controller and the controlled process are always combined into a closed loop either $x(t)$ or $y(t)$ is eliminated in a mathematical treatment. For eqn (9.1) this has already been done so that with given t, z or w only x is calculated as a function of time. With z constant for $t \geqslant 0$, $w=0$, and $s_1=0$, eqn (9.1) reduces to

$$z=r_{-1}\int_0^t x\,\mathrm{d}t+(r_0+s_0)x+(r_1+s_0T_1)\frac{\mathrm{d}x}{\mathrm{d}t}. \qquad (9.2)$$

Differentiation with z constant gives

$$r_{-1}x+(r_0+s_0)\frac{\mathrm{d}x}{\mathrm{d}t}+(r_1+s_0T_1)\frac{\mathrm{d}^2x}{\mathrm{d}t^2}=0. \qquad (9.3)$$

The solution for x is

$$x=C_1+C_2(\mathrm{e}^{p_1t}+\mathrm{e}^{p_2t}) \qquad (9.4)$$

where C_1, C_2, p_1 and p_2 are constants which depend on the geometry and the operation of the apparatus and

$$p_{12}=-\frac{r_0+s_0}{2(r_1+s_0T_1)}\pm\sqrt{\left[\frac{1}{4}\left(\frac{r_0+s_0}{r_1+s_0T_1}\right)^2-\frac{r_{-1}}{r_1+s_0T_1}\right]}.$$

Various cases have to be distinguished:

(1) $r_{-1}=0$. The controller contains no integral term. Then

$$p_1=-\frac{r_0+s_0}{r_1+s_0T_1} \quad \text{and} \quad p_2=0.$$

Taking into consideration the initial conditions $x=z/s_0$ for $t=0$ and

Fig. 9.3. The control loop for replication: model of error elimination and fixation of replication rate due to enzyme action.

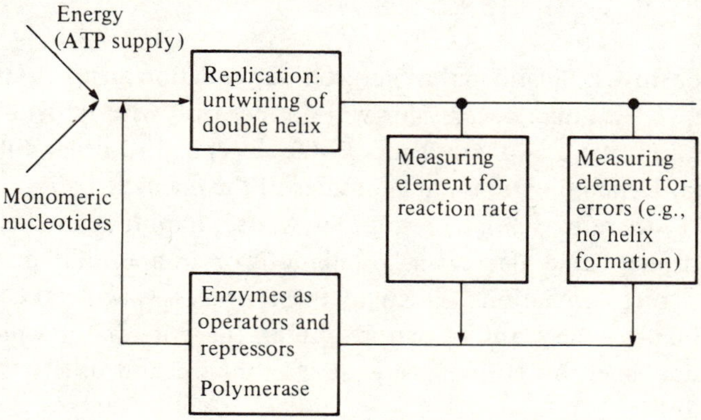

 $x = z/(r_0 + s_0)$ for $t = \infty$ yields

$$x = \frac{z}{r_0 + s_0} + z\left(\frac{1}{s_0} - \frac{1}{r_0 + s_0}\right) \exp\left(-\frac{r_0 + s_0}{r_1 + s_0 T_1} t\right). \tag{9.5}$$

We observe an exponential decay of the disturbance and a remaining control deviation $z/(r_0 + s_0)$ (Fig. 9.3).

(2) $r_{-1} \neq 0$. The controller has an integral-action component. Depending on the sign of the radicand we have exponentially decreasing oscillations

$$x = \frac{z}{s_0} \cdot \cos \sqrt{\left[\frac{r_{-1}}{r_1 + s_0 T_1} - \frac{1}{4}\left(\frac{r_0 + s_0}{r_1 + s_0 T_1}\right)^2\right]} t \exp\left(-\frac{r_0 + s_0}{2(r_1 + s_0 T_1)} t\right) \tag{9.6}$$

or an aperiodic decay

$$x = \frac{z}{s_0} \cdot \cosh \sqrt{\left[\frac{1}{4}\left(\frac{r_0 + s_0}{r_1 + s_0 T_1}\right)^2 - \frac{r_{-1}}{r_1 + s_0 T_1}\right]} t \times \exp\left(-\frac{r_0 + s_0}{2(r_1 + s_0 T_1} t\right). \tag{9.7}$$

The disturbance is completely removed.

With respect to biological cases, for eqn (9.5) these solutions mean that an external disturbance results in a permanent change (mutation) or a pathological degeneration. Eqns (9.6) and (9.7) imply that disturbances become ineffective after some time; oscillations may occur (see Fig. 9.3).

(3) With $r_{-1} = 0$, $w = 0$, $s_0 \neq 0$, $s_1 \neq 0$ and z constant, Eqn (9.1) becomes

$$z = (r_0 + s_0)x + (r_1 + s_1 + s_0 T_1)\frac{dx}{dt} + T_1 s_1 \frac{d^2 x}{dt^2}, \tag{9.8}$$

with the solution

$$x = \frac{z}{r_0 + s_0} - \frac{z}{r_0 + s_0} \cdot \cos \sqrt{\left[\frac{r_0 + s_0}{T_1 s_1} - \left(\frac{r_1 + s_1 + s_0 T_1}{2 T_1 s_1}\right)^2\right]} t \times \exp\left(-\frac{r_1 + s_1 + s_0 T_1}{2 T_1 s_1} t\right). \tag{9.9}$$

With a positive radicand in the characteristic equation an aperiodic decay to $x = z/(r_0 + s_0)$ may occur. This will not be dealt with in detail here.

(4) For $r_{-1} \neq 0$, $s_1 \neq 0$, $s_0 \neq 0$ and $T_1 \neq 0$ we get a third-order differential equation. Depending on the magnitudes of the coefficients r_0, r_1, r_{-1}, s_0, s_1 and T_1 there may arise oscillations whose amplitudes increase to a constant final value, decreasing oscillations or an aperiodic diminution of the control deviation. Which of these processes will take place is determined by the stability of the system, the criteria for which have been established by Hurwitz, Nyquist, Leonhard and others (see [76]) (Fig. 9.2*c*).

An integral term ($r_{-1} \neq 0$) tends to remove the disturbance completely (in cases (1) to (4)). Without an integral action effect ($r_{-1}=0$) a control deviation $z/(r_0+s_0)$ remains.

The solutions of the differential equation for closed control loops with a given disturbance z show oscillations or exponential decay down to zero or a remaining control deviation. The similarity of the solutions to those of irreversible thermodynamic processes is evident. Therefore we are justified in stating not only that the entropy concept of thermodynamics is analogous to that of information theory but also that the findings of irreversible thermodynamics are similar to those of cybernetics. In cybernetics the question of stability is decided on the basis of stability criteria. In comparing the results of the nonlinear thermodynamics of irreversible systems or the theory of dissipative systems with those of control engineering it should be noted that the latter considers a single control loop whereas thermodynamics deals with a large ensemble of systems of the same kind. As we find an enormous number of identical molecules in living beings the calculations for a single molecule may be transferred to an ensemble. We have recognized agreement between entropy in information theory and thermodynamics and also such a relation between nonlinear irreversible processes and cybernetics. Nevertheless, the ways of gaining this knowledge are quite different. The formal agreement points to fundamental connections between cybernetics and thermodynamic considerations on dissipative systems.

9.2 Biological control at the molecular level [22, 76]

In control loops at the molecular level the intermolecular forces and their assignment among neighbouring molecules play the part of the elements of the control loops and determine the coefficients r_0, r_1, r_{-1}, s_0, s_1, T_1, etc. It is a task of the future to determine the numerical values for these constants by experimental and theoretical research of the bonding forces between DNA and proteins and within these molecules. In replication the nucleotides and the basic molecule serving as template are the controlled system whose kinetic behaviour is characterized by the coefficients s_0 and s_1. The enzymes take the part of the controller (coefficients r_{-1}, r_0, r_1). Similar correspondences hold for transcription and translation. Likewise the description of a flow equilibrium in cybernetic terms is particularly illustrative. A discussion and mathematical formulation of all possibilities would be beyond the scope of this book. We will restrict ourselves to a few examples and will present no calculations.

 Fig. 9.3 shows the block diagram of the control loop for replication. The rate of replication is controlled and used as a control quantity. The formation of errors is denoted by $z(t)$. Special enzymes act as sensors and control the start, the termination and also the error elimination in this template polymerization. Similar control processes at the molecular level are found in protein synthesis. Again enzymes are responsible for starting and stopping transcription and translation. Additional control processes ensure that the genetic code is built up.

As we have seen, the establishment of a flow equilibrium as the basis for active life furnishes a closed control loop. Here the control quantity is the maintenance or desired decrease of entropy production. The results of control engineering calculations (eqns (9.1) to (9.9)) confirm the thermodynamic calculations of Prigogine, Glansdorff, Eigen and Ebeling [23, 25, 36, 84, 85].

The energy flow, i.e., the ATP supply for endothermic chemical processes in the cells, is also controlled by a feedback reaction which is equivalent to a controller. Again the aim is to maintain the flow equilibrium and to establish minimum entropy production. The decisive point for the maintenance of the flow equilibrium and the establishment of minimum entropy production is the determination of the rate of change of entropy production (on which the supply of ATP depends). In this context the derivative term $r_1\, \mathrm{d}x/\mathrm{d}t$ in the basic equations (9.1) and (9.8) is of particular importance. This 'derivative action' results in minimum entropy production in a very short time.

9.3 Macroscopic control processes in cells and organisms [22, 39, 69]

Besides control processes at the molecular level there are numerous others that occur at the macroscopic level and influence life processes. In assessing organ functions the motor and sensory complexes accessible to, and controlled by, consciousness are frequently overestimated while the role played by the control mechanisms which proceed automatically in the subconscious or excluding the nervous system is overlooked.

There are a number of complex control systems controlled by the vegetative nervous system or even by the endocrine circulation. We can also mention here the interplay of muscle groups in various coordinated movements which proceed almost unconsciously, such as walking, talking, cycling and writing. This involves complete control loops. Thus, e.g., the balancing mechanism in the ears determines control commands and uses these to actuate controlled countermotions of particular muscle groups via complicated connections in the cortex. The macroscopic biological control loops serve not only for constant value control, e.g., of the body temperature (Fig. 9.4) by influencing the combustion processes in the muscles, the supply of blood to the skin and evaporation at the body surface, but also for changes of other desired values which take place unconsciously.

Other control systems concern the blood pressure, blood sugar content, hydrogen ion concentration, sodium chloride content, the kidney and liver functions, etc. Even the control of the pupillary reflex would require an extensive description [22]. All these biological processes can be mentioned only briefly here, but this may suffice to show the complexity of life processes and the variety of practical solutions developed by organisms. In addition, there are nerve- or hormone-controlled functions for whole cell complexes, such as muscle contraction, assimilation by digestion, the formation of digestive ferments, etc. The next higher level is the interplay of numerous groups of muscles in complex motor processes. Even unconscious processes, such as heart action and breathing, are controlled by cortical centres which, in turn, are loosely connected with consciousness.

In complex biological control systems multiple control based on simultaneous information transmission via hormones and nerves as well as feedforward disturbance and networks with auxiliary control and correcting variables are found. This applies, above all, to the control of the most vital functions such as the heartbeat, blood pressure, body temperature, pH value, etc. We can suppose that at the molecular level the control loops have simpler structures.

Fig. 9.4. Complex control loop exemplified by the temperature control of a warm-blooded animal (model scheme).

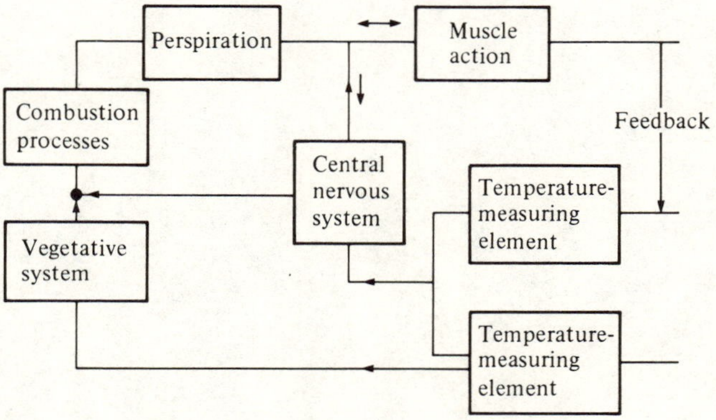

Besides the measurement of the control quantity x, the disturbance z (e.g., the deviation of the body temperature from the normal value) is found and fed into the controller. If different components in the nervous system and the hormonal circulation cooperate deviations can be optimally controlled. Sometimes this is done by influencing other organs without exceeding the desired value, for example in the case of perspiration. This implies a very fast response and a minimum disturbance. In the simplest case feedforward disturbance takes place according to the block diagram of Fig. 9.4. Also, auxiliary control variables x_a may be effective in addition to the control quantity proper, x. These might be, e.g., the mean body temperature and certain temperature values at the surface (for example, in the supercooling of individual limbs) (Fig. 9.4).

Finally, it should be pointed out that auxiliary correcting variables y_a may be employed. Thus supercooling of the body may give rise to an increased supply of blood, combustion processes in the muscles and involuntary muscular movements aimed at raising the body temperature (e.g., shivers). Exact calculations for such complex control systems are found in Oppelt's book [76].

10

Reflection and complementarity in biological processes

The concepts put forward in Chapters 10 and 11 represent my personal opinion and, being to some extent novel, are intended to stimulate discussions.

10.1 Introduction and definition of concepts

In this chapter I shall again refer to Engels' statement that in the future man would certainly be able to 'reduce' thinking to molecular and chemical motions in the brain, to which he added the question whether this would provide an exhaustive explanation of the nature of thinking [29, p. 264].

To a scientist the molecular and cellular processes are of importance and are primarily dealt with in research. The work of psychologists and philosophers, however, centres around the second part of the above statement. We ask the question as to where the two approaches are linked and whether we scientists can contribute to solving such a problem at all. Nonidealistic philosophy no longer doubts the coupling of mental processes in the brain and consciousness with materially recognizable chemical changes in neurons and biomolecules. The bridge between the two approaches is furnished by the Marxist theory of reflection [90], yet the concept of 'complementarity' must also be employed in the interpretation. What do these two concepts imply?

According to Klaus [59] reflection is the property of matter, e.g., of the brain, that reproduces external influences by means of internal changes and responds to them. Klaus interprets the concept of reflection in a very general manner. He extends these considerations even to material objects external to our consciousness. We find the origin of this opinion in the dialectic philosophy of Diderot in which 'sense and material reactions' are general properties of matter.

Following the example set by Niels Bohr complementarity will be used to mean that the aspects in question are two explanations of one and the same property that are complementary to each other. Further, one interpretation of the subject excludes the other. The best example is the explanation of light as waves in the one case and as corpuscles in the other. Sometimes an interpretation is based on a subjective influence of the experimenter or the experiment on the object. We do not take such a feedback of the thinking being to the conceived object into consideration. We consider our environment as a reality existing independently of us, outside our consciousness.

10.2 Reflection

I shall use the concept of reflection in a narrower sense here, referring primarily to the transmission of knowledge of the real world to our consciousness. The image originating in the brain, which is fixed in, and by means of, molecules, is of quite a different quality than reality. Thus the reproduction of acoustic vibrations perceived by our sense of hearing and stored in the brain via the nervous system may create in us a feeling of 'beautiful, of great artistic value, fascinating'. Physically speaking, a symphony represents a conglomeration of acoustic vibrations of the air; only the reflection in the brain makes it a masterpiece, the appropriate virtuosic sound generation being a prerequisite for creating the impression which enraptures us.

Complex analysis in mathematics, which is based on the definition of complex numbers, is another example of a logically-connected set of axioms which exists only in our brain and which can be illustrated by means of appropriate drawings in the real world. A work of poetry or prose fiction represents the description of an event or a development or sensations. The described action is not supposed to have happened or to happen in reality. (Consider fairy tales, science fiction, etc.)

In every case material carriers are required for all processes, within and without our consciousness. Several examples are given in Table 10.1. All the processes presented in Table 10.1 are connected by the fundamental concept of reflection. Only the material carriers are accessible to purely scientific cognition. Until now it has been impossible, and sheer speculation, to analyse the images in the cortex since their complexity cannot be grasped. Also unknown are possible processes at the molecular level that are released by the subconscious and which may influence evolution and affect regeneration phenomena.

Table 10.1. *Material carriers of processes in the consciousness*

Event	Material carrier	Reflection	Material carrier in the cortex	Consciousness
Mathematical knowledge	Formula drawing, ink, paper	→	Networks, biomolecules	Mathematical cognition, logical treatment
Written representation, poem, novel	Paper, printer's ink	→	Networks, biomolecules	Remembrance of certain events; sensations, judgements
Music	Score, magnetic tape, disk	→	Networks, biomolecules	Emotions, sensations
Human contacts	Networks, biomolecules	→	Networks, biomolecules	New knowledge, formation of opinions, stimulation to actions
Microworld	Atoms, molecules	→	Networks, biomolecules	Model concepts, transfer to already existing experience

10.3 Complementarity

In the biological context complementarity is the concept of the necessary completion of two or more aspects for understanding the whole. The original concept relates only to completion. Thus, two angles which together make 90° are called complementary and so are two colours which together give 'white'. According to Bohr the description of an event as a wave rules out treating the same phenomenon as corpuscles. In reality, however, there exists an electromagnetic excitation of space which can conveniently be described by a partial differential equation for the electric or magnetic field as a function of space and time; using the interference of waves the phenomena can be reduced to the microscopic level. The dilemma of 'waves versus corpuscles' seems to be due to the fact that the clear conceptions of a wave or a corpuscle have been transferred from the macroscopic world to the atomic level. We are sure that molecules and atoms exist in the real, recognizable world around us but the images we make of them (balls, nuclei with electrons surrounding them, waves, etc.) have been formed in our brain. This has been done by reflection via our sense organs. Hence they are models which are very useful in illustrating what is going on but cannot be credited with depicting reality in a perfect manner. For example, we can describe a corpuscle as a wave packet quenching itself by interference everywhere in space with the exception of a region of indeterminate size. The position of the wave packet is indeterminate if its velocity is measured very exactly. In this case comparatively few waves are required for the description. The more the region of nonquenching is limited the more waves of different wavelengths are needed. This requires greater velocity intervals because of the dispersion of the propagation velocity (indeterminacy of velocity). The exclusion principle mentioned here also plays a role in our point of view. However, we consider two concepts to be complementary if they necessarily go together and are indispensable for a complete description. Of course, in this sense the wave and the corpuscle are also complementary in the description of electromagnetic processes (Tables 10.2 and 10.3).

In the various sciences different emphasis is placed on the possible points of view. The two schemes illustrate respectively the role played by the concepts of reflection and complementarity and storage, control and processing at the molecular level.

It is considered possible that in rare cases organisms themselves may unconsciously intervene in their genotypic information storage due to the influence of selection pressure. Such a possibility would favour the formation of positive mutations and changes in the phenotype.

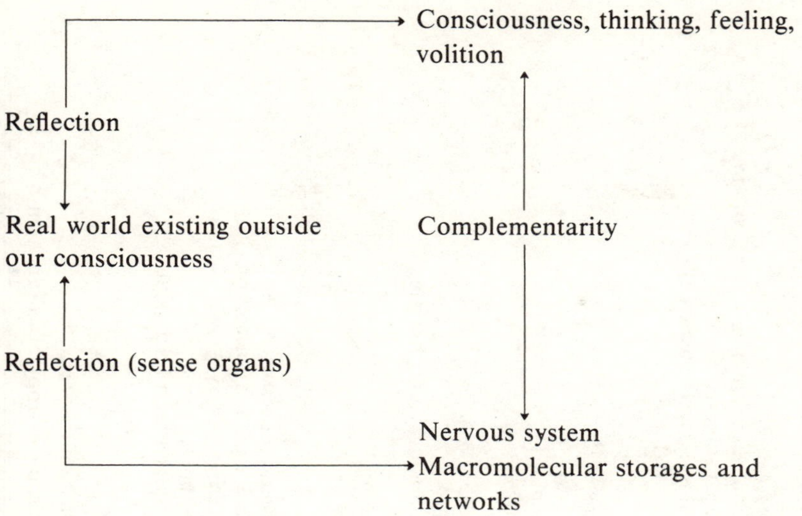
Consciousness, thinking, feeling, volition
Reflection
Real world existing outside our consciousness
Complementarity
Reflection (sense organs)
Nervous system
Macromolecular storages and networks

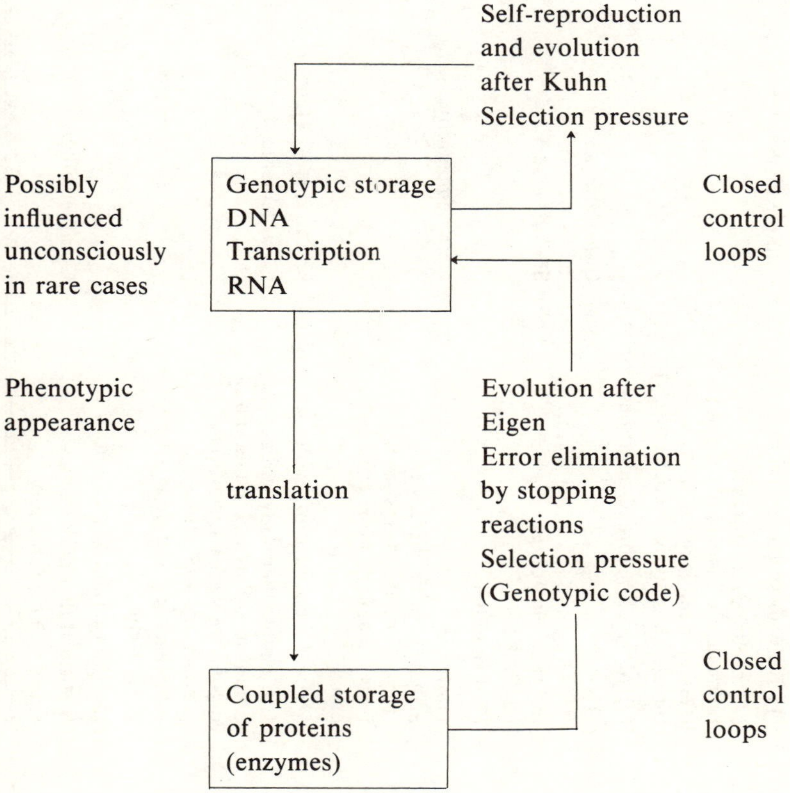
Self-reproduction and evolution after Kuhn
Selection pressure
Possibly influenced unconsciously in rare cases
Genotypic storage
DNA
Transcription
RNA
Closed control loops
Phenotypic appearance
translation
Evolution after Eigen
Error elimination by stopping reactions
Selection pressure
(Genotypic code)
Closed control loops
Coupled storage of proteins (enzymes)

Table 10.2. *Some examples of complementarity*

(1) Nervous system	← Complementarity →	Psyche
Macromolecular storages: biomolecules, molecular networks (Research subjects: biology, biophysics, biochemistry)		Consciousness, thinking, feeling, volition (Research subjects: psychology, psychomedicine)
(2) Molecular storages and their formation		
DNA RNA information storage legislative	← Complementarity → (as carriers of life processes)	Proteins (enzymes) functioning executive
(3) Computer		
hardware: arithmetic unit, addresses storage, (diodes, transistors, relays, magnetic tape, punched cards)	← Complementarity →	software: programming, algorithms, logic
(4) Self-reproduction, evolution, nonequilibria in open systems		
Irreversible thermodynamics: nonlinear differential equations, dissipative systems, entropy production, oscillations, establishment of a flow equilibrium	Complementarity ↔	Cybernetic processes: feedback equations of control engineering oscillations establishment of a final value
(5) Information and its entropy		
Statistical thermodynamics: entropy and probability	Complementarity ↔	Information theory: negentropy, entropy of information

Table 10.3. *Points of view in physics and biology*

Singularity		Complementarity
Physics		
Corpuscle	Wave	Wave and corpuscle
Classical physics	Classical physics	Quantum physics
Newton	Huygens	de Broglie
Democritus	Maxwell	Heisenberg
	Schrödinger	Bohr
Biology		
Pronouncedly materialist	Pronouncedly idealist	
Darwin	Mendel	Eigen
Oparin	Goethe	Engelhardt
Steinbuch		The author
Kuhn		

11

Development of the stock of biological information:

evolution and selection from a molecular aspect

11.1 Macromolecular storages in the frozen-in state and flow equilibria

Nobody now doubts that on the earth organisms have developed from primitive forms over a period of more than a thousand million years. The 'survival of the fittest' in connection with a selection pressure and a strong trend toward more variety and greater richness in forms is the governing factor. The genotypic storage has undergone a parallel development in three stages:

(1) The formation of coacervates, i.e., small protoplasm agglomerates without membranes and cell structure. Even in this chemical phase the tiny self-reproducing organic systems with primitive metabolism contributed to changing the environment. Here we should mention the precipitation of carbon dioxide and calcium oxide in the formation of organisms with calcareous shells and the generation of free oxygen with the beginning of photosynthesis in the atmosphere. Many scientists hold that the formation of carbohydrates with the help of chlorophyll (assimilation) was possible only after cells had been formed. A phase of life based on the oxidation of Fe^{II} to Fe^{III}, hydrogen sulphide and ammonia is supposed to have preceded photosynthesis. However, even these processes require the existence of free oxygen. Hence the question has to be thoroughly reconsidered whether a primary energy source due to the absorption of quantum radiation is responsible for maintaining feedback reactions and sustaining life processes. This first stage is called the chemical phase of life.

(2) The appearance of evolutionary processes. Unicellular organisms and, later, multicellular organisms appeared on the earth; the oxygen content of the atmosphere rose due to photosynthesis; and lime deposits of organic origin (shell limes, Jurassic limes, chalk) were formed. Thus

the originally existing large and excess amounts of atmospheric carbon dioxide were bound. The amount of carbon of organic origin contained in the limestone mountains is larger than the fossil carbon resources so that a high original carbon dioxide pressure (about 20 atm) can be assumed [47].

(3) The origination of warm-blooded animals, particularly birds and mammals. This implied such marked changes in the history of evolution that we may justly speak of a third phase.

A large amount of research work has been devoted to the development of polymeric biomolecules which ran parallel to that of living beings as a totality. We know that life as such is causally related to these information carriers, yet their origin is a controversial point.

Primitive forms of life possess information for the following processes:

(1) Autocatalysis for the initiation and maintenance of chemical reactions (energy uptake, metabolism, excretion);
(2) self-reduplication (replication) by means of template polymerization;
(3) maintenance of dissipative systems with minimum entropy production.

These controlled dynamic processes characterized by driving forces (affinities) and transformation of matter (flows) must be ensured for open thermodynamic systems with the uptake of free energy. The stock of genotypic information must be transmitted to the next generation without essential changes and loss of information. From the very beginning DNA and the proteins associated with it have been responsible for this. Therefore, the important point is that even at the beginning of life there were polymeric proteins and nucleic acids with their helices turned in a definite direction and with a base sequence carrying the basic information. We can estimate that a primary information content of 10^5–10^6 bits was sufficient. Various possible mechanisms of generation of these primary information carriers are under discussion. They are as follows:

(1) The formation of a definite base sequence completely by chance. Certain variations of sequence conformity are allowed mainly because not all purine and pyrimidine bases are carriers of information. If we assume the spontaneous generation of information carriers with an information content of 10^8 bits as a random process in the presence of high concentrations of purine and pyrimidine bases in a 'primaeval soup' containing sugar and phosphate then we find that there are 10^8 favourable arrangements of the side-group sequences as compared with the $4^{100\,000\,000}$ ones which are possible. The logarithm of the probability of a spontaneous generation is then found to be $\log P = -60 \times 10^6$. The negentropy

 for such an event is gigantic and the probability of $10^{-60\,000\,000}$ is so small that we can speak of an absolute impossibility. However, if we assume that about 10^5 bits of information are sufficient then $10^5/4.32$ or about 20 000 links are required for a primary protein chain. Here the probability of chain formation is $(1/20)^{10\,000} \approx 10^{-13\,000}$. Such a molecule would never be formed in any part of the universe by chance. Primary formation by chance would be like perpetual motion of the second kind, i.e., it would be incompatible with the second law of thermodynamics. (Perpetual motion of the first and second kind contravenes, respectively, the first and second laws of thermodynamics.) It is so improbable that no scientist assumes the formation of a primitive information carrier with a content of 10^5 bits in a 'primaeval soup'. Now and then the random formation is still interpreted as involving a primary information content of 10^3 bits. This gives a probability of about 10^{-400}. Simpson [99] and Baur [5] obtained a probability of 10^{-800} for the random primary generation of prebionts. This figure is also immensely small.

(2) The formation of these primary biomolecules by spontaneous generation. Many research workers (Eigen, Kaplan, Kuhn, Miller, Oparin and others [25, 26*a*, 55, 63, 71, 75]), as a result of the above, have adopted models of spontaneous generation which may be called orientated random processes. On the basis of ingenious assumptions and certain experiments, it is shown [26*c*] that by means of purposeful reactions in the sense meant by Darwin, and assuming an extremely favourable composition of the 'primaeval soup', self-reproduction, self-information and energy transfer may occur. This state involves definite concentrations of sugar, phosphate, amino acids, purine and pyrimidine bases as well as many other organic molecules. If the 'right' molecular chains with lengths of the order of 100 sequences are formed at first in the absence of templates then proteins with an enzymatic action may be generated. For the formation of a helical chain consisting of 20 nucleotides with the same turning direction throughout Kuhn [63] calculates a probability of 10^{-6}. The 'right' base sequences formed according to this assumption may serve as templates for further polymerization processes. Thus, single-stranded base sequences acting as information carriers can become predominant. Information carriers and contents originate simultaneously.

Further steps are necessary, namely, the enzymatic elimination of errors and controlled servo-mechanisms (second-order autocatalysis after Eigen [26*c*]). The information as well as its carrier must be checked. This is done by the formation of hypercycles, i.e., topological arrangements of quasi-species. The hypercycles are split up into compartments of different actions. Only then is the starting-point for further evolution and selection reached. The irreversible dynamics of nonlinear systems determines the direction in which the processes take place. The aim is known, i.e., in other words the systems which are subject to a random development are programmed in the sense of an algorithm such that from the consecutively-appearing random variants they always select the most favourable ones. It may be helpful in this context to consider game theory, a subdivision of cybernetics, which is extensively discussed in recent publications of Eigen [26*a*, 26*b*, 26*c*, 26*d*]. Starting from a random succession of nucleotides the desired (telenomic) messages may be characterized by a higher selection pressure (selection advantage [26*a*]). The probability of reduplicating the 'right' molecule by template polymerization is slightly higher than that for errors. The right base sequences in DNA are improved from generation to generation and the false ones are eliminated (Darwinism of molecules).

Let us consider an example of game theory. In a game at dice somebody wants to throw a five n times. On average it takes six throws to get the five. The die is left lying and with a second die a second five will be obtained after another six throws. On average the aim is reached after $6n$ throws ($P = 1/6n$ instead of $P = 1/6^n$). The latter case occurs if in the game n dice are thrown simultaneously. A similar mechanism might serve to perfect the genetic information storage. If the aim is known only one new nucleotide at the end of one chain is sufficient to store a new message with the information content of 8 bits. The formation of a new message which is hereditary takes place rarely, say after 1000 generations. The addition of either A or C or G or T or a false turn is possible (8 cases) and we reckon that one piece of information may be positive. All occasional false changes are normally eliminated again (one die is left lying in the above game!). Then, 8000 generations are necessary to stabilize a new positive message in the genotypic storage and 8000×10^8 generations are required to give a storage containing 10^8 bits of information. Further we have to consider that a newly-formed base sequence must occur simultaneously in several macromolecules carrying information (see Chapter 9). This makes the probable time of formation much longer. With a lifetime as short as 10^3 s for each generation it would take about 10^{20} s, i.e., about 10^{13} years for the stock of genetic information to develop. This is ten thousand times as long as the period of evolution estimated on the basis of geological and palaeontological measurements.

The probability of such a development is so small that even if evolution takes place on a million stars we have to suppose that we are the only living beings in the universe. This aspect is emphasized by the Soviet author Shklovsky [98]. He writes: 'Still, from a philosophical point of view, it would be interesting to ask the question: And what if we are alone in the universe? Only an astronomer who can really imagine the dimensions of the universe in space and time can fully comprehend the enormity of the idea that we should be alone and unique not only in space but also in time (for what are 350 years of technological development as compared with the 10 to 15 thousand million years of the evolution of the universe). If we were unique in the universe this would, in principle, mean the renaissance of a geocentric or even anthropocentric philosophy, although on an incomparably higher level than in the Middle Ages. The philosophers should think over these problems.'

On account of the importance of these considerations I shall look somewhat more closely at attempts to understand the individual steps in the formation of coacervates. Particular emphasis should be placed on the interpretation of these systems as molecular control loops with functional 'controllers' (enzymes) and controlled systems (DNA molecules). An important feature is the interplay of different processes which are directionally determined but which take place randomly with respect to time. In his excellent book *Probleme der Biogenese* (*Problems of Biogenesis*) [62] Körner gave a summary of the more than 2000 publications on this subject that had appeared by then. Of these, the most important were written by Urey [111], Kaplan [55], Oparin [75] and Miller [71] followed by fundamental model considerations put forward by Eigen [25, 26*b*, *c*, *d*] and Kuhn [63].

All theories that assume a primary origin of life on the earth or on an earth-like planet are based on the view that the formation of a prebiogenic structure undergoing reproduction, i.e., the formation of a thermodynamic nonequilibrium, must happen within a period of at most a thousand million years. Kuhn [63] discusses the conditions that have to be satisfied for such a dissipative system to be established. He assumes a number of individual steps which are considered possible but seem to be quite unlikely on closer examination. Extending the work of Glansdorff and Prigogine [36, 84] Eigen treats the maintenance of thermodynamic non-equilibria in dissipative systems subject to minimum entropy production. No doubt the above considerations, which describe dynamic molecular processes and correlate them with phenomena observed in control loops, concern the essentials of life.

Let us consider the 'start-up' or formation of dissipative systems with minimum entropy production as spontaneous generation, a process which, according to Eigen, certainly involved proteins and nucleic acids as active components. Following recent research results, spontaneous generation as understood in this sense may have taken place more than

10^9 years ago in a suitable environment on the earth. What remains unsatisfactory is the extremely high unlikelihood calculated for the spontaneous formation of such a dissipative system. Therefore, I shall now discuss possibilities for the formation of dissipative systems which can dispense with the assumption of perpetual motion of the second kind. We start from the observation that macromolecular systems can retain their primary structure without change in the base sequence and without entropy production for any period of time in a wide temperature range and in an inert environment. The systems in question are in frozen-in nonequilibria which means that unlikely configurations are retained. It is a fundamental consideration that macromolecules, such as DNA molecules, which represent a storage for the necessary basic information were present on the earth or got there even before dissipative systems became operative, i.e., before the beginning of dynamic entropy-producing processes. As environmental conditions sufficient to give rise to dynamic-flow equilibria developed on the earth the transition from the frozen-in, inactive molecular state to the dynamic-flow equilibrium with minimum entropy production took place. Hence the origin of life can be regarded as a kinetic phase transition. Its prerequisites are the presence of water, utilizable free energy and the existence of a number of basic substances, in particular phosphorus and nitrogen compounds [47]. The formation of the first stage of chemical life as a kinetic phase transition is compatible with the work of Kaplan, Kuhn, Oparin and others.

The subsequent evolution may have proceeded according to Kuhn's model or Eigen's considerations. But where does the basic information for the initiation of flow equilibria or the formation of simple control loops consisting of enzymes and DNA come from? Pertinent evidence was provided recently when organic matter was observed outside the earth and extraterrestrial organic substances and prebionts were detected in meteorites [111]. Thus, formaldehyde, formic acid, cyanides, formamide, methanol and a number of other low-molecular weight organic substances were found in cosmic dust and in interstellar space. Under the action of UV radiation these may also form amino acids. In this context papers of Winnewisser [119] and Johnson [53] must be mentioned. It can also be presumed [111] that organic molecules as carriers of information were involved in the formation of the primaeval earth and gave rise to prebiological chemical evolution. Lovell [65] asserts that very primitive organisms existed in the dust of the cold solar nebula from which the sun and the planets originated. This assumption is also extensively discussed in [47]. Pflug [81] deals with the question as to whether the time necessary for the formation of prebiological forms, i.e., the time for the formation of a primary stock of information, could have been long enough or whether one should not assume life to be older than the earth. This view is supported by Claus, Nagy and Europa [13] who detected spore-like organic entities in the Orgueil meteorite. Convincing evidence for the penetration into the geosphere of extraterrestrial information carriers and the subsequent start-up of dissipative systems is provided by the work of Hoyle & Wickramasinghe [49*c*]. While in the present book statistical considerations back these concepts up and refute the theory of random origin, Hoyle & Wickramasinghe base their conclusions on astronomical observations.

In my previous publication [47], I discussed the question of how these macromolecular systems can survive the conditions prevailing in space. Since primary cosmic radiation destroys only the atomic nucleus, and the intensity of UV radiation is low at large distances from fixed stars, macromolecules as carriers of information may be preserved over periods of more than 10^9 years. Of course, this concept finally leads to a steady-state theory as postulated by Bondi [11] and Hoyle [49*a*]. It also requires a supply of biomolecules from planets carrying life to interstellar space. At present most physicists are convinced that about 10 000 million years ago a 'big bang' marked the beginning of our world. The red shift of the spectral lines of distant fixed stars, the 3 K radiation and the helium concentration in the universe are evidence for this assumption. This led to the theory of an expanding universe. On the other hand, a very weak exchange effect between 'gravitons' and 'photons' analogous to the Compton effect can also explain these observations. If we agree to the theory of 'big bang' then an original creation of dissipative systems as carriers of biological information by chance needs to have happened only once on some fixed star in the last 10^{10} years. The probability of this event is about 10^{12} times greater than for the formation of life on every planet. In a 'steady-state' theory it is useless to ask for the first carrier of this fundamental information and to assume a perpetual motion at the beginning of life. The concept of transferring material carriers of life processes between the stars was confirmed by Hoyle [49*b*] from an astronomical point of view.

Finally, the reader should be warned not to consider extraterrestrial macromolecules which store certain information as living systems in space. The beginning of life on the earth as discussed here is not a unique event if informative macromolecules are assumed to exist everywhere. Every planet belonging to a fixed star and possessing chemical and physical properties which allow a biosphere to develop will be a carrier of life processes. At present the earth is the only planet in our solar system fulfilling these conditions. In many millions of years time this might also be true for Venus if the sun continues to cool down.

11.2 The chemical phase of life

If water, carbon dioxide, nitrogen, sulphur, phosphorus, potassium, calcium and some further elements that are required in small amounts are present, then frozen-in macromolecular states can make transitions to dissipative systems in the temperature range between 0 and 100 °C according to the concept elaborated in this book. Additional decisive factors for this kinetic phase transition, which should take place in the aqueous phase, are the absence of catalyst poisons, especially hydrogen sulphide, sulphur dioxide, ammonia, arsenic compounds and heavy metal compounds in high concentrations, and protection from harmful radiations [47]. At the very beginning living systems may exist as drops in the gel state (Gordon) without membranes. Membranes are formed later (they are important for chemical coherence, metabolism and safety). It is in this phase of life that the weathering of silicates and the formation of sediments of calcium carbonates on an organic basis begins. For this a period of several thousand million years must be assumed. This means that the period of the beginning of life is longer than the phase which is characterized by evolution. The chemical period is thoroughly described by Körner [62] and many other authors.

11.3 Evolution

The epoch characterized by the gradual chemical shaping of the environment through the help of primitive organisms changes almost imperceptibly to the era of evolution. As an organism assumes its shape, i.e., its phenotype, the related molecular storage is being formed. The origination of any class, family or species of animals or plants presupposes the further development of this storage. A well-defined and identically-structured basic storage is assigned to all living beings. As more and more features are included the storage becomes increasingly complicated. Different models have been suggested for evolution in which development is based on predetermined randomness and the assumption of a certain selection pressure. Nonlinear thermodynamics proves the elimination of unfavourable (nonsurviving) systems. Eigen [25, 26*a*] considers evolution as a complex process based on the nucleic acid–enzyme correlation. The error frequency of 1 : 100 in the replication of DNA that he assumes is largely compensated for by the elimination of errors during protein synthesis.

Only the mechanism of varying the genotypic storage very slightly enables organisms to develop into higher forms. In thermodynamic terms evolution is characterized by the relation

$$\frac{dS}{dt} < 0$$

for an open living system. In contrast to this, if the flow equilibrium is retained,

$$\frac{dS}{dt} = 0.$$

It should be noted that information is transmitted only in the direction DNA → RNA → protein. During evolution, higher-structured entities are formed in events which are unlikely but sometimes necessarily occur. Normally, the formation of the genotypic storage of subsequent generations is determined by Mendel's laws, which are discussed in standard works of biology.

As evolution progresses, i.e., as the genotypic storage becomes more and more complex, there is an increasing differentiation of the organ functions, a transition from simple to complex forms accompanied by an adaptation to the changing environmental conditions. Palaeontologists have obtained evidence for the evolution of primitive forms as early as the Cambrian. The evolution of species is among the most solid fundamentals of biology.

Mutations, which result in the appearance of new complicated organs, go beyond the boundaries of species. The associated quantity–quality transition may take place within comparatively few generations. This provides an explanation as to why in many cases intermediate forms cannot be detected. The progress of the stock of genotypic information is coupled with the development of the phenotype. The basic information is preserved in this process. New information, i.e., new qualities, is grafted onto this basis and stored in additional genes. In embryogeny the primitive rudiments become a reality. The principle of evolution has priority over economic requirements. For example, warm-blooded animals are inferior to cold-blooded ones with respect to energy turnover but not with respect to vitality and richness in experience. Therefore, warm-blooded animals represent the higher stage. The genotypic storage which has materialized in macromolecules has a history extending over a thousand million years. The development of DNA as the carrier of information is accompanied by the formation of phenotypes. Table 11.1 gives the timetable for evolution.

Evolution fulfils a biological target function, which consists of the transition from simple to complex, from poor in experience to rich in

	Event	Time in seconds (order of magnitude) before the present
(1)	Origin of life (formation of the first flow equilibria or dissipative systems with reduplication and autocatalysis)	10^{17}
(2)	Chemical era in biological development (flow equilibria, self-reproduction, template polymerization, photosynthesis (?), CO_2 degradation, calcium bonding, development of the genotypic storage up to about 10^8 bits)	10^{17} to 2×10^{16}
(3)	Era of evolution (development of micro-organisms and primitive animal and plant forms, particularly those bearing calcareous shells; extinction of whole classes of animals and plants; origination of vertebrates)	2×10^{16} to the present
(4)	Warm-blooded animals (increase of the genotypic storage to about 10^{10} bits, extinction of many species)	2×10^{15}
(5)	Man (increase of the genotypic storage to about 10^{12} bits)	10^{13}
	Technological age	10^{10}
	Average life-span of human individual (during this time the memory is formed and experience is acquired, the storage capacity totalling about 10^6–10^7 bits)	2×10^9

experience, from instinctive to 'consciously' controlled. Living beings become more adaptive, more versatile, more mentally flexible and active. In primitive forms the actions are controlled by instincts; as the mental power increases, more and more learning processes and actions induced by the will take place.

There are several principles governing life.

(1) *The principle of economy.* A minimum entropy production in the sense of Prigogine and a long constancy of thermodynamic nonequilibria is to be accomplished with a minimum consumption of free energy. This leads to the formation of open thermodynamic systems having ambient temperature, in other words, to the development of cold-blooded animals. Biological substances are to be involved in the organic cycle as long as possible. By the absorption of light quanta plants generate organic substances which, when taken up by animals as food, are degraded down

 to low-molecular weight units and again synthesized to polymers specific to a given species or energy carrier. The natural end for herbivorous animals is to be eaten by carnivores. Thus they are a link in the food chain plant → herbivorous animal → carnivore. In this way the biological substance enters other living beings so as to make the life of microorganisms also possible. In general, however, the degradation proceeds only to a point where the remains can enter other organisms. Nitrogenous substances, e.g., are degraded only down to ammonia or the nitrates.

An apparent waste in biological processes contrasts with this principle of economy. Only a small fraction of all seeds and sperms gives rise to new living beings. Yet this 'waste' heightens the probability of mutations and thus results in increasing the number of individual peculiarities. The tendency to increase the number of forms and species has priority over the principle of economy. Maintaining the biological equilibrium necessarily requires degradation processes which are kept going by heterotrophic organisms. (Heterotrophic organisms are all those whose life processes are made possible by organic substances and which are incapable of photosynthesis.)

(2) *The principle of evolution.* As could be shown by several examples, evolution, i.e., the formation of new species by mutation and selection, has priority over economy in the organic kingdom. Evolution leads to the formation of organisms with the greatest wealth of experience, with the highest rate of gaining this and to the storage of further information. Only in rare cases is this knowledge which is acquired by learning included in the genotypic storage. This requires gathering the same experience over many generations and cannot happen within a few generations as was supposed by Lamarck. The 'survival of the fittest' means the survival of the most developed species rather than of the economically most favoured ones. If the principle of economy were true the microorganisms, the bacteria, would have got the upper hand of the higher-developed organisms and extinguished them. Hereditary changes of the phenotype can be brought about only by changes in the sequences of the individual-specific DNA. Man-made changes in the base sequence are called genetic engineering.

(3) *Ecology.* Ecology deals with the coexistence of many organisms and their interdependences. The concept of ecology can be explained by the principle of the preservation of biological equilibrium and the suppression of processes disturbing this equilibrium. Technology as developed by man and the transformations of the environment taking place in civilization are frequently opposed to the basic laws of ecology. Yet nature takes vengeance and counteracts the technological activities

of man. The processes involved are based on several ecological principles:

(1) A buffer action in response to excessive environmental changes. Thus, the emission of large amounts of carbon dioxide into the atmosphere gives rise to the dissolution of larger amounts of carbon dioxide in the oceans. Also, the high concentration of organic and inorganic waste in bodies of water causes counteractions. Plant growth increases until eutrophication is reached. In other words, the oversupply of nutrients results in instability and the eventual extinction of life.

(2) Encroachment on the living conditions of individual species leads to their extinction or the unbridled proliferation of other organisms. It means a disturbance of the ecologically determined food chains. Thus, the large-scale killing of whales in the Antarctic Ocean has already resulted in an enormous increase of the crustaceans which serve as food for those mammals.

(3) The development of a flora and fauna of interdependent species and forms gives rise to the multiplicity of life and the variety of experiences for all living beings that we know today. The wealth of shapes and colours constitutes the beauty of the earth. It should be one of the most important tasks of man in the future not to destroy any more but to restore these values. Therefore, it is of prime importance that present-day scientists should study the laws of ecology.

We have seen in the present book how the biological macromolecule has life processes assigned to it in an all-embracing manner. The subject matter of the last two chapters is the correlation between molecular processes, life and experience. Obviously, it is not enough to investigate only the composition and structure of molecules but it is just as important to comprehend the correlations with all their implications. However, this would require considering the life processes at the various biological levels, which is beyond the scope of this book. It is hoped, however, that the reader will feel the need to continue to read textbooks of biology. In so doing he will recognize that though molecular biology is the fundamental part of biology it cannot claim to interpret all life processes from just one point of view. Therefore, it seems appropriate to conclude this chapter by quoting Tyuchev's lines which serve as the motto of Wolkenstein's [121] book:

> Nature is not what you think,
> it is not a cast, a soulless face.
> It has a soul, in it is freedom,
> in it is love, it speaks to us.

12

Concluding theses

(1) All life processes including thinking, feeling and volition are connected with material, recognizable chemical processes in macromolecular systems.

(2) Only macromolecules have the variability required for different stable chemical compounds. They cannot be destroyed by simple conformational changes (i.e., rotational or translational motions).

(3) Template polymerization produces identical macromolecules of exactly-defined sizes and base sequences of the constituent monomeric units.

(4) Macromolecular systems obey the laws of thermodynamics without exception. In particular, the theory of flow equilibria allows the existence of dissipative systems far from thermodynamic equilibrium with continuing entropy production.

(5) Thermodynamic nonequilibria are frozen-in in many polymers. By means of template polymerization such systems reproduce without consuming free energy. There is a tendency for a frozen-in nonequilibrium to be maintained over a rather large range of temperatures and under different environmental conditions.

(6) The basic equation

$$\mathrm{d}S = 1/T\,\mathrm{d}U + \mathrm{d}S_\mathrm{i} - 1/T\,\mathrm{d}F_\mathrm{e} \quad \text{(cf. eqns 5.21, 5.22)}$$

governs the life processes. In the dissipative system the entropy change $\mathrm{d}S_\mathrm{i}$ caused by irreversible processes is compensated by the supply of external free energy $\mathrm{d}F_\mathrm{e}$. Free energy must be continually supplied to a living system to balance the entropy produced that flows into the environment ($\mathrm{d}S_\mathrm{e} > 0$) against the decrease in entropy due to growth, information increase and evolution ($\mathrm{d}S_\mathrm{i} < 0$).

(7) The relations $dS_L \leq 0$ and $dS_D > 0$ are equally important. For the open system of a living organism (subscript L) entropy decreases with growing differentiation, e.g., in the form of growth and learning processes. In the case of death (subscript D), which coincides with the end of feedback reactions, the system which has been far from thermodynamic equilibrium tends to the latter. The living system and the environment (subscript E) together constitute a closed thermodynamic system in which the second law of thermodynamics is valid without restriction: $dS_L + dS_E > 0$. Living systems may change the thermal contact with their environment consciously or unconsciously.

(8) Thermodynamic nonequilibrium is maintained by feedback reactions in the sense of Prigogine [84, 85] and Eigen [25, 26]. These are automatic control processes in all living cells and their structural units. They also include autocatalysis. The mathematical apparatus of cybernetics explains the molecular events in the cells as well as many other processes at the higher levels of life. The feedback reactions are causally related to replication, protein synthesis and the self-reproduction of larger units. They are in accord with the equations of irreversible thermodynamics with respect to possible oscillations, permanent deviations and stability.

(9) Macromolecular systems are the carriers of all biological information. On the one hand this refers to the genotypic information, the basic elements of which can be assumed to be the same for all living beings. It includes the directives for autocatalysis, self-reproduction, feedback reactions and the energy balance. At an early stage of biological development a separation took place as a result of which only plants include the capability of photosynthesis in their genotypic information storage. On the other hand biological information means the neural information (memory) which is gained in the course of life and also stored in macromolecular systems.

(10) Owing to the correlation of information ↔ material fixation in the molecule, the basic equation of information theory,

$$H = -\sum_i P_i \log P_i, \quad \text{with} \sum_i P_i = 1,$$

which corresponds to the thermodynamic relation

$$S = k \ln P,$$

acquires more than formal significance. Increasing information content is associated with the formation of more complex, i.e., less likely molecules. This material assignment requires decreasing entropy, or rising negentropy, as the complexity of the molecules increases.

(11) Information can be stored in macromolecules in different ways. Considering possible errors it is improbable that the sequence of purine and pyrimidine bases starting from the beginning of a very long molecule serves to mark information. It seems more favourable to assign definite polyads (diads, triads, etc.) to definite messages inside the molecule. Moreover, the combination of several polyads into complex systems in a manner similar to the formation of sentences in a language is possible for this purpose.

(12) Possible errors are reduced by repeated storage (redundancy).

(13) Living systems and robots show formal resemblances, yet the latter are not capable of active development independent of external help and they are incapable of self-reproduction. They can operate following only the stored instructions and information. Automata have no freedom of the will.

(14) If we assume that the most primitive forms of life (originating coacervates) had information contents of more than 10^3 bits, then the random spontaneous generation of these coacervates in an appropriate environment would be perpetual motion of the second kind. If, however, spontaneous generation is considered as a phase transition, i.e., as the formation of a dynamic dissipative system from a frozen-in informative nonequilibrium, then the origination of life on planets is possible.

(15) To account for the development of very simple dissipative systems into very highly ordered complex states it is necessary to assume chains of orientated random processes. A period of about three thousand million years would be too short for absolutely random evolution.

(16) The material components realized in macromolecular biological systems reflect the life qualities experience, learning, feeling, thinking and volition in the sense of dialectical materialism.

(17) The real and recognizable world around us, its reflection through receptors and nerves in the macromolecular systems of the cortex and its treatment on a mental level are interrelated as shown.

Reflection receptors	↔	Material carriers of life phenomena (DNA, proteins)
↕		↕
Real world existing outside our consciousness		Complementarity
↕		↕
Reflection sphere of experience	↔	Thinking, feeling, volition (psyche, memory, consciousness)

(18) Information storage in macromolecules is similar to that of other material carriers of information: ink and paper for mathematical formulae, magnetic tape for music and graphic representation for poems, prose works and communications. The information content (that which is carried) as such is complementary to this and of a completely different quality.

(19) If definite environmental conditions are given [47], life will necessarily begin with the transformation of frozen-in carriers of information into dissipative systems and further into coacervates, then into microorganisms and finally into multicellular organisms. We are not the only living beings in the universe.

REFERENCES

1 Aurich, H. (1973). *Laboratorium des Lebens.* Leipzig: Urania-Verlag.
2 Autorenkollektiv (1976). *Kleine Enzyklopädie Leben.* Leipzig: VEB Bibliographisches Institut.
3 Autorenkollektiv (1961–70). *Der Mensch und das Leben. Eine Enzyklopädie der biologischen Wissenschaften.* (Translated from French. Original title: *La vie et l'homme.*) Leipzig: Urania-Verlag.
4 Barrell, B. G., Air, G. M. & Hutchinson, C. A. (1976). Overlapping genes in bacteriophage ϕX 174. *Nature,* **264**, 34.
5 Baur, F. (1969). Kosmische Aspekte der organischen Evolution. *Naturwissenschaftliche Rundschau,* **22**, 167.
6 Bertalanffy, L. von (1958). *Biophysik des Fließgleichgewichts.* Braunschweig: Vieweg.
7 Bertalanffy, L. von (1965). Die Biophysik offener Systeme. *Naturwissenschaftliche Rundschau,* **18**, Heft 12.
8 Bertalanffy, L. von (1973). *General System Theory.* Harmondsworth, Middlesex: Penguin Books Ltd.
9 Bielka, H. (1973). *Molekulare Biologie der Zelle.* Jena: VEB Gustav Fischer Verlag.
10 Blumenfeld, L. A. (1977). *Probleme der molekularen Biophysik.* Berlin: Akademie-Verlag.
11 Bondi, H. (1960). *Cosmology,* 2nd edn. Cambridge: Cambridge University Press.
12 Braitenberg, V. (1973). *Gehirngespinste, Neuroanatomie für kybernetisch Interessierte.* Berlin and Heidelberg: Springer-Verlag.
13 Claus, G., Nagy, B. & Europa, D. L. (1961). A microbiological examination of some carbonaceous chondrites. *Nature,* **192**, 594. (1963). Further observations on the properties of the 'organized elements' in carbonaceous chondrites. *Annals of the New York Academy of Sciences,* **108**, 580.
14 Crick, F. H. C. (1970). Central dogma of molecular biology. *Nature,* **227**, 561.
15 Crick, F. H. C., Barnett, L. S., Brenner, S. & Watts-Tobin, R. I. (1961). General nature of the Genetic Code for proteins. *Nature,* **192**, 1227.
16 Crick, F. H. C., Griffith, I. S. & Orgel, L. E. (1960). On protein synthesis. In *Biosynthesis of macromolecules,* p. 209 (Soviet edn). Moscow: Foreign Literature Publishers.
17 Crick, F. H. C. (1966). The Genetic Code – yesterday, today and tomorrow. *Cold Spring Harbor Symposia,* **31**, 3.

18 Delbrück, M. & Stent, G. (1968). Der Mechanismus der DNS-Reduplikation. In *Chemische Grundlagen der Vererbung*. Moscow: Staatsverlag der UdSSR.

19 Dingman, W. & Sporn, M. B. (1964). Molecular theories of memory. *Science*, **144**, 26.

20 Doty, P. (1960). Configurations of biologically important macromolecules in solutions. In *Modern Problems of Biophysics*, 1 (Soviet edn). Moscow: Foreign Literature Publishers.

21 Driesch, H. (1920). *Leib und Seele*. Leipzig: E. Reinecke.

22 Drischel, H. & Tiedt, N. (Eds.). (1971). *Biokybernetik*, vols. 1 and 2. Jena: VEB Gustav Fischer Verlag.

23 Ebeling, W. (1976). *Strukturbildung bei irreversiblen Prozessen*. Leipzig: BSB B. G. Teubner Verlagsgesellschaft.

24 Eccles, J. C. (1964). *The Physiology of Synapses*. Berlin, Göttingen, Heidelberg and New York: Springer-Verlag.

25 Eigen, M. (1971). Selforganization of matter and the evolution of biological macromolecules. *Die Naturwissenschaften*, **58**, 465.

26*a* Eigen, M. (1973). Interview: Manfred Eigen's Theorie über die Evolution biologischer Makromoleküle. *Umschau*, **73**, 420.

26*b* Eigen, M. (1980). Das Urgen. *Nova Acta Leopoldina*, **52**, 5.

26*c* Eigen, M., Gardiner, W., Schuster, P. & Winkler-Oswatitsch, R. (1981). Ursprung der genetischen Information. *Spektrum der Wissenschaft*, Juniheft, 36.

26*d* Eigen, M. & Schuster, P. (1979). *The Hypercycle*. Heidelberg: Springer-Verlag.

26*e* Eigen, M. & Winkler-Oswatitsch, R. (1981). Transfer RNA an early gene? *Die Naturwissenschaften*, **68**, 282.

27 Elsasser, W. (1958). *The Physical Foundation of Biology*. Oxford: Oxford University Press.

28 Engelhardt, W. A. (1969). Das Problem des Lebens in der modernen Naturwissenschaft. *Sowjetwissenschaft*, **22**, 734. (1972). Hierarchien in biologischen Systemen. Plenary lecture at the Conference of the Biophysical Society of the GDR, Luisenthal.

29 Engels, F. (1952). *Dialektik der Natur*. Berlin: Dietz-Verlag.

30 Fey, P. (1963). *Informationstheorie*. Berlin: Akademie-Verlag.

31 Försterlin, H. D., Kuhn, H. & Tews, K. H. (1972). Computermodell zur Bildung selbstorganisierter Systeme. *Angewandte Chemie*, **84**, 862.

32 Fuchs-Kittowski, K. (1972). *Information, ihre Speicherung und Verarbeitung in biomolekularen Systemen*, II. *Kühlungsborner Kolloquium*. Berlin: Akademie-Verlag.

33 Fuchs-Kittowski, K. (1976). *Probleme des Determinismus und der Kybernetik in der molekularen Biologie*. Jena: VEB Gustav Fischer Verlag.

34 Geissler, E. (Ed.). (1972). *DNS – Schlüssel des Lebens*, 2nd revised edn. Berlin: Akademie-Verlag.

35 Geissler, E. (1972). *Taschenlexikon Molekularbiologie*. Leipzig: VEB Bibliographisches Institut.

36 Glansdorff, P. & Prigogine, I. (1971). *Thermodynamic Theory of Structure, Stability and Fluctuation*. New York and London: Wiley Interscience Publishers.

37 Glees, P. (1971). *Das menschliche Gehirn*. Stuttgart: Hippokrates Verlag.

38 Groot, S. R. de (1952). *Thermodynamics of Irreversible Processes*. Amsterdam: North Holland Publishing Company.

39 Glaser, E. (1971). *Einführung in die Biophysik*. Jena: VEB Gustav Fischer Verlag.

40 Harbers, E. (1969). *Nukleinsäuren.* Stuttgart: Georg Thieme.

41 Hassenstein, B. (1972). *Biologische Kybernetik.* Jena: VEB Gustav Fischer Verlag.

42 Henze, E. (1967). *Einführung in die Informationstheorie.* Berlin: VEB Deutscher Verlag der Wissenschaften.

43 Hodgkin, A. L. (1964). *The Conduction of the Nervous Impulse.* Liverpool: Liverpool University Press.

44 Hofmann, E. (1972). *Dynamische Biochemie,* vols. 1, 2, 3, 4. Berlin: Akademie-Verlag.

45 Holzmüller, W. (1961). *Physik der Kunststoffe.* Berlin: Akademie-Verlag.

46 Holzmüller, W. (1961, 1966). *Technische Physik,* vol. 1. Berlin: VEB Verlag Technik. Vol. 2. BSB B.G. Teubner Verlagsgesellschaft.

47 Holzmüller, W. (1977). *Unsere Umwelt – ihre Entwicklung und Erhaltung.* Leipzig: BSB B.G. Teubner Verlagsgesellschaft. 3rd edn, 1984.

48 Holzmüller, W. (1978). Molecular mobility, deformation and relaxation processes in polymers. *Advances in Polymer Science,* **26**, 1.

49*a* Hoyle, F. (1968). *Die Natur des Universums.* Cologne: Verlag Kiepenheuer und Witsch.

49*b* Hoyle, F. & Wickramasinghe, C. (1981). Where microbes boldly went. *New Scientist,* **1**, 412.

49*c* Hoyle, F. & Wickramasinghe, C. (1981). (1982). Wie das Leben zuf die Erde kam. *Bild der Wissenschaft,* **1**, 38.

50 Huxley, J. (1964). *Entfaltung des Lebens.* Stuttgart: Fischer-Verlag.

51 Hyden, H. (1967). *The Neuron.* Amsterdam: Elsevier Publishing Company.

52 Jacob, F. & Monod, J. (1961). Genetic regulatory mechanisms in the synthesis of proteins. *Journal of Molecular Biology,* **3**, 318.

53 Johnson, F. M. (1972). Interstellar Matter II: Diffuse interstellar lines and porphyrins. *Annals of the New York Academy of Sciences,* **187**, 186.

54 Jung, F. (1977). Zweite Grundthese der Molekulargenetik wankt. *Wissenschaft und Fortschritt,* **27**, 146.

55 Kaplan, R. V. (1972). *Der Ursprung des Lebens.* Stuggart: Georg Thieme.

56 Karlson, P. (1970). *Kurzes Lehrbuch der Biochemie.* Stuttgart: Georg Thieme.

57 Katz, B. (1974). *Muskel und Synapse.* Stuttgart: Georg Thieme.

58 Keller, A. (1957). A note on single crystals in polymers: Evidence for a folded chain configuration. *Philosophical Magazine,* **2**, 1171.

59 Klaus, G. & Buhr, M. (1971). *Philosophisches Wörterbuch.* Leipzig: VEB Bibliographischses Institut.

60 Klix, F. (1971). *Information und Verhalten. Kybernetische Aspekte der organischen Informationsvermittlung.* Berlin: VEB Deutscher Verlag der Wissenschaften.

61 Kornberg, A. (1960). Biological synthesis of DNA. *Science,* **131**, 1503. (1961). *Enzymatic Synthesis of DNA.* New York: John Wiley & Sons.

62 Körner, U. (1978). *Probleme der Biogenese.* Jena: VEB Gustav Fischer Verlag.

63 Kuhn, H. (1972). Selbstorganisation molekularer Systeme und die Evolution des genetischen Apparates. *Angewandte Chemie,* **84**, 838.

64 Libbert, E. (1977). *Kompendium der allgemeinen Biologie.* Jena: VEB Gustav Fischer Verlag.

65 Lovell, B. (1964). *Neue Wege zur Erforschung des Weltraumes.* Göttingen: Verlag Vandenhoeck and Ruprecht.

66 Lorente de Nó (1962). *Circulation of Impulses and Memory in Molecular Specificity and Biological Memory.* Cambridge, Massachusetts: Francis O. Schmitt.

67 Matthies, H. (1972). Die intrazelluläre Regulation der interneuralen Konnektivität – Eine Grundlage von Lern-und Gedächtnisprozessen. In *Biocybernetics* IV. Jena: VEB Gustav Fischer Verlag.
68 Matthies, H. & Biesold, D. (1977). *Neurobiologie*. Jena: VEB Gustav Fischer Verlag.
69 Mittelstädt, H. (1961). *Regelungsvorgänge in lebenden Wesen*. Munich: R. Oldenbourg Verlag.
70 Monod, J. (1973). *Zufall und Notwendigkeit. Philosophische Fragen der modernen Biologie*, 5th edn. Munich: R. Piper.
71 Miller, S. L. & Orgel, I. F. (1974). *The Origin of Life on the Earth*. New York.
72 Neidhardt, P. (1964). *Informationstheorie und automatische Informationsverarbeitung*. Berlin: VEB Verlag Technik.
73 Neurath, H. (1961). *The Proteins*. New York: Academic Press.
74 Noll, H., Staehelin, T. & Wettstein, F. (1963). Ribosomal aggregates engaged in protein synthesis, ergosome breakdown and messenger ribonucleic acid transport. *Nature*, **198**, 632.
75 Oparin, A. Y. (1963). *Das Leben, seine Natur, Herkunft und Entwicklung*. Jena: VEB Gustav Fischer Verlag. (1968). *Origin of Life. Genesis and Evolutionary Development of Life*. New York: Academic Press. (1975). Evolution of general ideas on the origin of life in the years 1924–1974. *Izvestija Akademii Nauk SSSR, Serija Biologičeskaja*, **1**, 5.
76 Oppelt, W. (1960). *Kleines Handbuch technischer Regelvorgänge*. Weinheim: Verlag Chemie.
77 Pauling, L. & Corey, R. (1953). Fundamental dimensions of polypeptide chains. *Proceedings of the Royal Society, B*, **141**, 10.
78 Pavlov, I. P. (1926). *Die höchste Nerventätigkeit (das Verhalten) von Tieren*, 3rd edn. Munich: J. F. Bergmann.
79 Perutz, M. F. (1966). *Proteins and Nucleic Acids. Structure and Function*. Amsterdam: North Holland Publishing Company.
80 Pfleiderer, G. & Jeckel, D. (1957). Individuelle Milchsäurehydrogenasen bei verschiedenen Säugetieren. *Biochemische Zeitschrift*, **329**, 370.
81 Pflug, H. D. (1971). Ist das Leben älter als die Erde? *Umschau*, **71**, 619.
82 Pierce, J. R. (1965). *Phänomene der Kommunikation*. Düsseldorf: Econ-Verlag.
83 Preuss, H. (1966). *Quantenchemie für Chemiker*. Leipzig: Akademische Verlagsgesellschaft Geest and Portig K.-G.
84 Prigogine, I. (1947). *Etude Thermodynamique des Procès Irreversibles*. Liège: Desoer.
85 Prigogine, I. & Rice, S. A. (1975). *Advances in Chemical Physics*, **32**.
86 Raisbeck, G. (1970). *Informationstheorie*. Berlin: Akademie-Verlag.
87 Rapoport, S. M. (1969). *Medizinische Biochemie*. Berlin: VEB Verlag Volk und Gesundheit.
88 Raven, C. (1961). *Oogenesis, The Storage of Development Information*. Oxford: Pergamon Press.
89 Reinbothe, H. & Krauss, G. J. (1982). *Entstehung und molekulare Evolution des Lebens*. Jena: VEB Gustav Fischer Verlag.
90 Rosenthal, S., Rosenthal, H. A., Fuchs-Kittowski, K. & Rapoport, S. M. (1977). *Biochemische und molekularbiologische Grundlage der Widerspiegelung*, IV. *Kühlungsborner Kolloquium*. Berlin: Akademie-Verlag.
91 Rüdiger, B. (1977). *Mechanismen und Lokalisierbarkeit der Informationsspeicherung im Zentralnervensystem*, IV. *Kühlungsborner Kolloquium*. Berlin: Akademie-Verlag.

92 Sato, K. (1971). Activity and interactivity of physiological systems. *Cybernetics*, **9**, 45. Jena: VEB Gustav Fischer Verlag. (1969). On the interactivity and recovery function in physiological systems. *Kybernetik*, **6**, 146. Berlin: Springer Verlag.

93 Schadé, J. P. (1973). *Die Funktion des Nervensystems.* Jena: VEB Gustav Fischer Verlag.

94 Schlegel, H. (1969). *Allgemeine Mikrobiologie.* Stuttgart: Georg Thieme.

95 Schrödinger, E. (1944). *What is life?* Cambridge: Cambridge University Press.

96 Shannon, C. E. (1976). *Mathematische Grundlagen der Informationstheorie.* Munich and Wien: R. Oldenbourg Verlag.

97 Sheraga, H. A. (1961). *Protein Structure.* New York: Academic Press.

98 Shklovsky, I. S. (1973). *Sowjetwissenschaft, Gesellschaftswissenschaftliche Beiträge*, **26**, 762.

99 Simpson, G. (1968). Kosmische Aspekte der organischen Evolution. *Naturwissenschaftliche Rundschau*, **21**, 425.

100 Slater, J. C. (1963). *Quantum Theory of Molecules and Solids*, vol. 1. London: McGraw-Hill.

101 Steinbuch, K. (1965). *Automat und Mensch.* Berlin and Göttingen: Springer-Verlag. (1972). Mensch und Maschine. *Nova Acta Leopoldina*, **37**, 206.

102 Straub, B. (1972). *Enzyme, Moleküle, Lebenserscheinungen.* Leipzig: Akademische Verlagsgesellschaft Geest and Portig K.-G.

103 Stuart, H. A. (1952). *Die Struktur des freien Moleküls.* Berlin, Göttingen and Heidelberg: Springer-Verlag.

104 Stuart, H. A. (1959). Kristallisationsbedingungen und morphologische Strukturen bei Hochpolymeren. *Kolloid-Zeitschrift*, **165**, 3.

105 Tembrock, G. (1971). *Biokommunikation. Informationsübertragung im biologischen Bereich*, WTB. Berlin: Akademie-Verlag. (1972). Informationsübertragung bei der organischen Kommunikation. *Nova Acta Leopoldina*, **37**, 206.

106*a* Tosi, C. (1971). The randomness in copolymers with penultimate mechanism of copolymerization. *Die makromolekulare Chemie*, **150**, 199.

106*b* Tosi, C., Corradini, P., Valvassori, A. & Ciampelli, F. (1969). Infrared determination of sequence distribution and randomness in copolymers. *Journal of Polymer Science*, **22**, 1085.

107 Träger, L. (1975). *Einführung in die Molekularbiologie.* Jena: VEB Gustav Fischer Verlag.

108 Trintscher, K. S. (1967). *Biologie und Information.* Leipzig: BSB B.G. Teubner Verlagsgesellschaft.

109 Ungar, G. (1968). Molecular mechanisms in learning. *Perspectives in Biology and Medicine*, **11**, 217.

110 Unsöld, A. (1978). Strukturen des Denkens. *Naturwissenschaftliche Rundschau*, **31**, Heft 8.

111 Urey, H. C. (1962). Origin of life forms in carbonaceous chondrites. *Nature*, **193**, 1119.

112*a* Voe, H. de & Tinoco, J. (1962). The hypochromism of helical polynucleotides. *Journal of Molecular Biology*, **4**, 518.

112*b* Vogel, G. & Angermann, H. (1979). *Taschenbuch der Biologie*, I und II. Jena: VEB Gustav Fischer Verlag.

113 Völz, H. (1976). Informationsspeicherung in Natur und Technik. *Wissenschaft und Fortschritt*, **26**, 242. (1977). Grenzen und Möglichkeiten der Informationsspeicherung. *Zeitschrift für elektrische Informations- und Energietechnik*, **7**, 470. (1977). Entwicklung und Stand der Informationsspeicherung. *Nachrichtentechnik*, **27**, 138.

114 Völz, H. (1981). *Elektronik*, 3rd edn. Berlin: Akademie-Verlag.

115 Watson, J. D. (1965). *Molecular Biology of the Gene.* New York: W. A. Benjamin.

116 Watson, J. D. & Crick, F. H. C. (1953). Genetic implications of the structure of desoxyribonucleic acid. *Nature*, **171**, 964.

117 Weizsäcker, C. F. von. (1972). *Einheit der Natur.* Munich: Carl Hanser Verlag. (1975). Die philosophische Interpretation der modernen Physik. *Nova Acta Leopoldina*, **37**, 207.

118 Wiener, N. (1963). *Kybernetik.* Düsseldorf and Wien: Econ-Verlag.

119 Winnewisser, G. (1975). Interstellar molecules in the galaxy. *Naturwissenschaften*, **62**, 200.

120 Wolf, G. (1974). *Neurobiologie*, WTB. Berlin: Akademie-Verlag.

121 Wolkenstein, M. W. (1969). *Moleküle und Leben.* Leipzig: VEB Georg Thieme.

122 Zilliken, F. & Abdallah, K. (1973). *Molekularbiologische Grundlagen des Kurz- und Langzeitgedächtnisses.* Stuttgart, New York: F. K. Schattauer-Verlag.

INDEX

Bibliothèques
Université d'Ottawa
Echéance

Libraries
University of Ottawa
Date Due

19 NOV. 1985
NOV 15 1985